BestMasters

Springer awards „Best Masters" to the best master's theses which have been completed at renowned universities in Germany, Austria, and Switzerland.

The studies received highest marks and were recommended for publication by supervisors. They address current issues from various fields of research in natural sciences, psychology, technology, and economics.

The series addresses practitioners as well as scientists and, in particular, offers guidance for early stage researchers.

Vladimir Herdt

Complete Symbolic Simulation of SystemC Models

Efficient Formal Verification of Finite Non-Terminating Programs

With a Preface by Prof. Dr. Rolf Drechsler

 Springer Vieweg

Vladimir Herdt
Bremen, Germany

BestMasters
ISBN 978-3-658-12679-7   ISBN 978-3-658-12680-3 (eBook)
DOI 10.1007/978-3-658-12680-3

Library of Congress Control Number: 2016930259

Springer Vieweg

Printed on acid-free paper

Springer Vieweg is a brand of Springer Fachmedien Wiesbaden
Springer Fachmedien Wiesbaden is part of Springer Science+Business Media
(www.springer.com)

Preface

Electronic systems consist of a large number of interacting hardware- and software components. Only the hardware often is assembled of more than a billion transistors. To cope with
this increasing complexity system description languages, which allow modeling at high level
of abstraction, have been introduced and are focus of current research. They facilitate architectural exploration as well as Hardware/Software Co-Design. SystemC has become the de-facto
standard for modeling at the system level.

The SystemC model serves as reference for subsequent development steps. Errors in a SystemC model are very critical, since they will propagate and become very costly. Thus, develop-
ing verification methods for SystemC is of very high importance. Existing formal verification
approaches at lower abstraction levels are already very sophisticated and require a profound
technical understanding to be further improved.

This book makes a significant contribution to the ongoing research activities. It is based on
the master thesis of Vladimir Herdt, which he has written as a student in the Group of Computer Architecture (AGRA) at the University of Bremen, Germany.

Compared to existing approaches, the proposed method works more efficiently and thus can
verify much larger (non-terminating) systems. It works by extending symbolic simulation with
stateful exploration and efficiently handling repeating behavior in the presence of symbolic
values. Besides the carefully formulated theoretical presentation, the proposed method and
additional optimizations have been implemented and evaluated on relevant examples. The com-
prehensive evaluation clearly shows the advantages of the proposed approach.

The significance of the obtained scientific results is further confirmed by the publication "Verifying SystemC using Stateful Symbolic Simulation", which is based on the thesis and has been
accepted for presentation at Design Automation Conference (DAC) 2015, San Francisco, USA
– the most renowned conference for electronic system design.

I hope you will enjoy reading this book.

October 2015

Rolf Drechsler
drechsler@uni-bremen.de

Contents

List of Figures

List of Tables

List of Algorithms

List of Listings

Acronyms

AVPE	Assertion Violation Preserving Exploration
BMC	Bounded Model Checking
DPE	Deadlock Preserving Exploration
DPOR	Dynamic Partial Order Reduction
ESM	Explicit Structural Matching
ESS	Exact Symbolic Subsumption
IB	Induction Base
IS	Induction Step
IH	Induction Hypothesis
POR	Partial Order Reduction
SAT	Boolean Satisfiability Problem
SBC	Solver Based Comparison
SBC-EQ	Solver Based Comparison with Equality Heuristics
SBC-CV	Solver Based Comparison with Subsumption/Coverage Heuristics
SDPOR	Stateful Dynamic Partial Order Reduction
SMT	Satisfiability Modulo Theories
SPOR	Static Partial Order Reduction
SSPOR	Stateful Static Partial Order Reduction
SSR	State Subsumption Reduction

1. Introduction

1.1. Motivation

System-on-Chips (SoC) are ubiquitous nowadays. They integrate an increasingly large number of hardware and software components on a single chip. The development of such complex systems is very challenging, especially within todays tight time-to-market constraints. To cope with this rising complexity, the level of abstraction is raised beyond the Register Transfer Level (RTL) to the Electronic System Level (ESL) [BMP07]. A higher level of abstraction allows for easier exploration of design alternatives and facilitates the parallel development and integration of hardware and software components. SystemC has become the de-facto standard in the ESL design [GD10]. Technically, it is a C++ library providing standardized components that facilitate the development of electronic systems. The structure of a SystemC design is described in terms of modules, ports, interfaces and channels, while the behaviour is described in processes. An event-driven simulation kernel is provided which executes the processes non-preemptively in an arbitrary order. SystemC designs are developed using a behavioral/algorithmic style in combination with abstract communication based on Transaction Level Modeling (TLM). Since the SystemC design serves as initial (executable) reference for subsequent development steps, ensuring its correctness becomes crucial, especially for safety critical systems, as undetected errors will propagate and become very costly.

1.2. Verification of SystemC Designs

1.2.1. Simulation-Based Methods

The straightforward approach to SystemC verification is to simulate the design with the reference kernel provided by Accellera [Acc12]. For a given set of input values, the produced output values are compared with the expected output values to detect errors. To enhance simulation-based verification, several approaches have been proposed to bring Assertion-Based Verification (ABV) [Fos09] into the SystemC context [EEH07; FP09; TV10]. Different extensions of the Property Specification Language (PSL) [IEE05] have been introduced to specify (TLM) properties in accordance with the SystemC simulation semantics [Tab+08; Eck+06; PF08]. This extensions enable to specify complex (temporal) properties and check them during the simulation. The idea is to translate a property to monitoring logic, then embed it into the design in form of simple source-code assertions and check for their violations on the fly.

However, these approaches only offer a limited verification coverage, since they explore only a single sequence of processes, called a *scheduling*. Due to the non-deterministic simulation semantics of SystemC, many different schedulings may exist. To increase the verification coverage by simulation, different methods have been proposed to exhaustively explore all valid schedulings. They employ variants of Static and Dynamic Partial Order Reduction (POR) [God96; FG05], denoted as SPOR and DPOR respectively, to reduce the number of redundant schedulings.

The approaches [Hel+06; HMM09] start by exploring a random scheduling. Based on the observed process interactions, alternate schedulings are explored. A method based on DPOR is employed to avoid the exploration of equivalent schedulings. [KGG08] first employs a static analysis to infer transition dependencies. A transition of a process is essentially a list of statements that is executed without interruption, due to the non-preemptive scheduling semantics of the SystemC simulation kernel. The statically pre-computed informations are queried at runtime by a DPOR based method to calculate process dependencies based on the dynamically explored paths. [BK10] uses model checking techniques to compute precise process dependencies during a preliminary static analysis. Then the SystemC program is transformed into a C++ program, which can be normally compiled and executed. The SystemC scheduler and the statically pre-computed race conditions are embedded into the C++ program to explore all relevant schedulings effectively.

Albeit providing better coverage, these methods still need representative inputs and thus cannot guarantee the absence of errors. Formal methods are required to ensure correctness of a design. They can either prove mathematically that a specific property holds or provide a counter example that shows a violation.

1.2.2. Formal Methods

Formal verification methods for SystemC need to consider all possible inputs and all possible process interleavings in order to ensure correctness. However, its object oriented nature and sophisticated simulation semantics make formal verification of SystemC very challenging [Var07]. Early efforts in formal verification of SystemC designs, for example [MMM05; KEP06; Tra+07], have very limited scalability or do not model the SystemC simulation semantics thoroughly [KS05]. Among the more recent approaches, the following show promising results and can currently be considered the state of the art in the formal verification of SystemC[1]:

- STATE, first proposed in [HFG08], translates SystemC designs to timed automata. With STATE it is not possible to verify properties on SystemC designs directly. Instead, they have to be formulated on the automata and can then be checked using the UPPAAL model checker [Amn+01]. Although this approach is complete, it is potentially inefficient as UPPAAL is not geared toward dealing with large input spaces, as shown by the experimental evaluation in [Le+13].

- SCIVER [GLD10] translates SystemC designs into sequential C models first. Temporal properties using an extension of PSL [Tab+08] can be formulated and integrated into the C model during generation. Then, C model checkers can be applied to check for assertion violations. High-level induction on the generated C model has been proposed to achieve completeness and efficiency. However, no dedicated techniques to prune redundant scheduling sequences are provided.

- Kratos [CNR13; Cim+11] translates SystemC designs into threaded C models. Then, the ESST algorithm is employed, which combines an explicit scheduler and symbolic lazy abstraction [Cim+10]. POR techniques are also integrated into the explicit scheduler [CNR11]. For property specification, simple C assertions are supported. Although this

---

[1]Part of this description already appeared in [Le+13].

approach is complete, its potentially slow abstraction refinements may become a performance bottleneck.

- SDSS [Cho+12] formalizes the semantics of SystemC designs in terms of Kripke structures. Then, Bounded Model Checking (BMC) and induction can be applied in a similar manner as in SCIVER. The main difference is that the scheduler is not involved in the encoding of SDSS. It is rather explicitly executed to generate an SMT formula that covers the whole state space. Still, no dedicated techniques to handle equivalent scheduling sequences are supported. This limitation is removed in [CCH13] by employing a so called *symbolic POR*. The symbolic simulation engine is used to precisely capture transition independencies, which are then used to dynamically infer process dependencies. A heuristic based on [KWG09] is employed to prune redundant scheduling interleavings during the state space exploration. Furthermore symbolic state merging is used to combine multiple execution paths to reduce the number of explored paths. Although this recent extension is quite efficient (an advantageous comparison with Kratos is also provided), it is incomplete and can only prove assertions in terminating SystemC programs.

- [Le+13] proposes an Intermediate Verification Language (IVL) for SystemC. Basically the (SystemC) IVL is a compact language that can be much more easily processed and analyzed, yet it is sufficient to describe the behaviour of SystemC programs. The intention of the IVL is to clearly separate the tasks of a SystemC front-end and a SystemC back-end. Furthermore a symbolic simulator called SISSI has been proposed to verify IVL programs. In particular SISSI can discover or prove the absence of assertion violations and other types of errors such as division by zero or memory access violation[2]. SISSI employs *symbolic simulation*, a combination of complete scheduling exploration and symbolic execution [Kin76; CDE08], to exhaustively explore the state space. SPOR and DPOR are used to prune redundant schedulings. Experimental evaluation in [Le+13] shows that SISSI outperforms the other listed approaches (with the exception of SDSS, which is not available for evaluation). However, a major limitation of SISSI is its incompleteness, due to the core algorithm that performs a stateless search, i.e. no record of visited states is kept. Consequently, it cannot avoid re-exploration of already visited states, and thus can only be applied to models that either terminate or contain bugs, and is unable to verify assertions over cyclic state spaces.

1.3. Thesis Goal: Complete Symbolic Simulation

As mentioned earlier, formal verification methods are essential to guarantee the correctness of SystemC designs. In contrast to simulation-based methods, they can prove the absence of errors. The goal of this thesis is to extend the symbolic simulator SISSI with a stateful search that is capable of detecting cycles. The combination of a stateful search with the existing symbolic simulation overcomes the limitation of SISSI in verifying assertions over cycle state spaces of non-terminating SystemC programs. This combined approach will be called *complete symbolic simulation*. To the best of the authors knowledge such a combination has not yet been considered in the context of SystemC. The following challenges arise when attempting such a combination:

1. POR has already been employed with great success in SISSI. However, a naive combination of the existing methods with stateful model checking can potentially lead to

---

[2]These other types of errors can also be formulated in form of assertions.

unsoundness, i.e. assertion violations can be missed. This is due to the (transition/action) *ignoring problem*, which refers to a situation, where a relevant transition is not explored. Furthermore, in the case of DPOR, additional care needs to be taken when calculating process dependencies, in order to preserve soundness. This situation, where a relevant dependency is missed, will be referred to as the *missing dependency problem* in this thesis. This problem can arise in both acyclic and cyclic state spaces.

2. Similarly, the efficient combination of the existing symbolic execution with a stateful search is challenging. A stateful search requires a process called *state matching* to decide whether a state has already been visited to avoid re-exploration. Symbolic execution stores and manipulates symbolic expressions, which represent sets of concrete values. Therefore, *symbolic state matching* is required which involves comparison of different symbolic expressions. The problem is that symbolic expressions can be represented in many (actually infinitely) different ways, e.g. the expressions $a + a$ and $2 * a$ are structurally different (hence not equal) but semantically equivalent.

3. Furthermore, state equivalence is an unnecessary strong requirement for symbolic state matching. The reason is as follows. A symbolic state represents a set of concrete states. If the set of concrete states represented by a symbolic state s_2 contains the set of concrete states represented by a symbolic state s_1, it is not necessary to explore s_1 if s_2 has already been explored. Symbolic state matching using equivalence would miss this reduction opportunity.

4. Finally, POR and symbolic state matching can be employed together as complementary reduction techniques. Nonetheless, additional care needs to be taken to avoid unsoundness.

These challenges require non-trivial solutions, including rigorous formalization to prove their soundness.

1.4. Thesis Contributions and Organization

The contributions of this thesis attempt to provide solutions for the described challenges. They are summarized as follows:

- A stateful search with SPOR is implemented. A *cycle proviso* is integrated to solve the ignoring problem. More details on the ignoring problem in the context of SystemC, a theoretical discussion of the correctness of this approach and the actual implementation are presented in Chapter 3.

- The stateless DPOR algorithm is extended to a stateful algorithm in three steps. The first step provides support for acyclic state spaces. This extension is adapted from the existing work [Yan+08] to accommodate the SystemC specific simulation semantics. In the second step a novel extension is proposed to handle cyclic state spaces. Finally, in the third step, a cycle proviso is integrated to solve the *ignoring problem* in general. These extensions will be further discussed in Chapter 4.

- State Subsumption Reduction (SSR) is proposed in Chapter 5 as a powerful symbolic state matching technique. The general idea is to match states by *subsumption* instead of equality in a stateful search. A symbolic state s_1 is said to be *subsumed* by s_2, or equivalently s_2 *subsumes* s_1, if the set of states represented by s_1 is a subset of the set of states represented by s_2. Thus, it is unnecessary to explore s_1 when s_2 already has been explored. The correctness of this approach will be established, this includes the introduction of a novel cycle proviso in combination with POR, and an exact algorithm provided. The algorithm is capable of detecting every possible subsumption of symbolic values, but is computationally very expensive.

- Different heuristics for more efficient state subsumption detection are proposed in Chapter 6 as alternatives to the exact algorithm. They include explicit and solver-based methods with different configurations, to balance between precision and runtime overhead. The goal is to increase the overall scalability, by spending less time in state matching.

The necessary theoretical background for the contributions will be presented in Chapter 2. The overall *complete symbolic simulation* approach of this thesis is evaluated in Chapter 7. As part of the experiments it is compared against the latest version of Kratos [CNR13]. The results show the efficacy and potential of the proposed *complete symbolic simulation*. The conclusion and some ideas for future work are presented in Chapter 8. The Appendix (Chapter A) contains proofs and some additional informations, that will be referenced from within the thesis.

Related Work The related work with respect to SystemC verification has already been discussed in Section 1.2. Here only related work directly connected to the contributions of this thesis is considered.

The *ignoring problem* has first been identified in [Val89]. The solution is to impose an additional condition called a *cycle proviso/condition*. Different cycle provisos have been proposed since then in combination with different search strategies [God96; HGP92; BH05; BLL06; EP10]. In this thesis the cycle proviso proposed in [EP10] is used to solve the ignoring problem in the SystemC context, when using POR in combination with a basic stateful search. However, this proviso is unsuitable when POR is combined with SSR, as it is too restrictive. Thus a weaker proviso is proposed in this thesis, to preserve safety properties when combining POR and SSR. To the best of the authors knowledge, such a proviso has not yet been proposed.

DPOR has been first introduced in [FG05]. It works by exploring an arbitrary sequence of transitions until completion and dynamically tracks relevant runtime informations to compute dependencies. These are used to detect and explore alternate relevant interleavings of transitions. DPOR supports stateless exploration of acyclic state spaces. A stateful extension is non trivial and can lead to unsoundness (under-approximation by missing relevant transitions) when implemented naively. As already mentioned, this problem will be referred to as *missing dependency problem* in this thesis. [Yan+08] have extended DPOR to support stateful exploration of terminating multi-threaded programs. Thus, their extension does not support cyclic state spaces. Another stateful DPOR extension has been proposed in [YWY06]. According to the authors of [YWY06], their algorithm supports cyclic state spaces, but it is unclear whether the ignoring problem has been solved. Due to the special simulation semantics of SystemC, these approaches are not directly applicable. To the best of the authors knowledge, DPOR has only been provided in a stateless version so far [KGG08; Hel+06; HMM09] in the context of SystemC. The stateful DPOR (SDPOR) implementation proposed in this thesis combines the stateless DPOR algorithm for SystemC, as presented in [KGG08], with the approach presented in [Yan+08] to support acyclic state spaces. To support cyclic state spaces a novel method is proposed, instead of adapting and incorporating the approach proposed in [YWY06]. Compared to [YWY06], the proposed approach is more lightweight.

The subsumption detection between symbolic values, which is part of the implemented SSR, is based on a method that appeared in [APV06] in the context of Java symbolic model checking. They combine the method with additional abstractions, that result in under-approximations of the normal behaviour, thus their combined approach is used for bug hunting.

2. Preliminaries

This chapter provides necessary background regarding SystemC, the IVL, symbolic execution, POR and basic definitions for state transition system, which can be used to model finite SystemC programs. Some parts of the SystemC and IVL description in Section 2.1 and Section 2.2 already appeared in [Le+13].

2.1. SystemC

SystemC is implemented as a C++ class library. It includes an event-driven simulation kernel and provides common building blocks to facilitate the development of embedded systems.

The structure of a SystemC design is described with ports and modules, whereas the behavior is described in processes which are triggered by events and communicate through channels. SystemC provides three types of processes with SC_THREAD being the most general type, i.e. the other two can be modeled by using SC_THREAD. A process gains the runnable status when one or more events of its sensitivity list have been notified. The simulation kernel selects one of the runnable processes and executes this process non-preemptively. The kernel receives the control back if the process has finished its execution or blocks itself by executing a context switch. A context switch is either one of the function calls *wait(event)*, *wait(time)*, *suspend(process)*. They will be briefly discussed in the following. Basically SystemC offers the following variants of *wait* and *notify* for event-based synchronization [GLD10; IEE11]:

- *wait(event)* blocks the current process until the notification of the event.

- *notify(event)* performs an *immediate notification* of the event. Processes waiting on this event become immediately runnable in this *delta cycle*.

- *notify(event, delay)* performs a *timed notification* of the event. It is called a *delta notification* if the delay is zero. In this case the notification will be performed in the next *delta phase*, thus a process waiting for the event becomes runnable in the next *delta cycle*.

- *wait(delay)* blocks the current process for the specified amount of time units. This operation can be equivalently rewritten as the following block { sc_event e; notify(e, delay); wait(e); }, where *e* is a unique event. Thus the *wait(delay)* variant will not be further considered in the following.

More informations on *immediate-*, *delta-* and *timed*-notifications will be presented in the next section which covers the simulation semantics of SystemC.

Additionally, the *suspend(process)* and *resume(process)* functions can be used for synchronization. The former immediately marks a process as suspended. A suspended process is not runnable. The *resume* function unmarks the process again. It is a form of delta notification, thus its effects are postponed until the next *delta* phase of the simulation. Suspend and resume are complementary to event-based synchronization. Thus a process can be suspended and waiting for an event at the same time. In order to become runnable again, the process has to be resumed again and the event has to be notified.

2.1.1. Simulation Semantics

The execution of a SystemC program consists of two main steps: an *elaboration* phase is followed by a *simulation* phase. During *elaboration* modules are instantiated, ports and channels are bound and processes registered to the simulation kernel. Basically elaboration prepares the following simulation. It ends by a call to the *sc_start* function. An optional maximum simulation time can be specified. The simulation kernel of SystemC takes over and executes the registered processes. Basically simulation consists of five different phases which are executed one after another [IEE11].

1. *Initialization*: First the *update* phase as defined in step 3 is executed, but without proceeding to the subsequent *delta notify* phase. Then all registered processes, which have not been marked otherwise, will be made runnable. Finally the *delta notify* phase as defined in step 4 is carried out. In this case it will always proceed to the *evaluation* phase.

2. *Evaluation*: This phase can be considered the main phase of the simulation. While the set of runnable processes is not empty an arbitrary process will be selected and executed or resumed (in case the process had been interrupted). The order in which processes are executed is arbitrary but deterministic[1]. Since process execution is not preemptive, a process will continue until it terminates, executes a wait statement or suspends itself. In either case the executed process will not be runnable. Immediate notifications can be issued during process execution to make other waiting process runnable in this evaluation phase. Once no more process is runnable, simulation proceeds to the *update* phase.

3. *Update*: Updates of channels are performed and removed. These updates have been requested during the evaluation phase or the elaboration phase, if the *update* phase is executed (once) as part of the initialization phase. The evaluation phase together with the update phase corresponds to a *delta cycle* of the simulation.

4. *Delta Notify*: Delta notifications are performed and removed. These have been issued in either one of the preceding phases. Processes sensitive on the notification are made runnable. If at least one runnable process exists at the end of this phase, or this phase has been called (once) from the *initialization* phase, simulation continues with step 2.

5. *Timed Notification*: If there are timed notifications, the simulation time is advanced to the earliest one of them. If the simulation exceeds the optionally specified maximum time, then the simulation is finished. Else all notifications at this time are performed and removed. Processes sensitive on these notifications are made runnable. If at least one runnable process exists at the end of this phase, simulation continues with step 2. Else the simulation is finished.

After simulation the remaining statements after *sc_start* will be executed. This phase is often denoted as *post-processing* or *cleanup*. Optionally *sc_start* can be called again, thus resuming the simulation. In this case the *initialization* phase will not be called. The simulation will directly continue with the *evaluation* phase. An overview of the different phases and transitions between them is shown in Figure 2.1.

---

[1]If the same implementation of the simulation kernel is used to simulate the same SystemC program with the same inputs, then the process order shall remain the same.

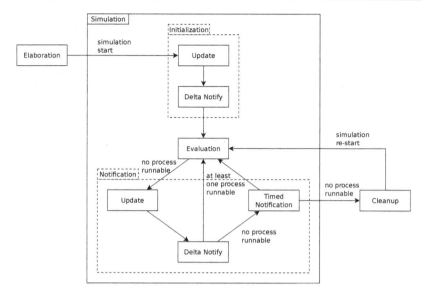

Figure 2.1.: Execution phases of a SystemC program. The *notification* phase is defined in this thesis as additional phase to group the *update*, *delta notify* and *timed notification* phases.

The non-determinism in the *evaluation* phase of the simulation, due to multiple process scheduling alternatives, is one of the reasons that give rise to the state explosion problem when it comes to the verification of SystemC programs. In order to assure that no failure is missed during simulation, it becomes necessary[2] to explore all relevant scheduling orders.

Remark. In the following the *update*, *delta notify* and *timed notification* phases will often be grouped as *notification* phase. Thus if simulation moves from the *evaluation* phase to the *update* phase, it will be said that the simulation is in the *notification* phase.

The interested reader is referred to [GD10; Gro02; Bla+09] or the IEEE standard [IEE11] for more details on SystemC.

2.2. SystemC Intermediate Verification Language

The SystemC Intermediate Verification Language (IVL) [Le+13] has been defined with the purpose of simplifying the verification process of SystemC programs, by separating it into two independent steps. The idea is that first a *front-end* converts a SystemC program into an IVL program, which is then verified by a separate *back-end*. The IVL has been designed to be compact and easily manageable but at the same time powerful enough to allow the translation of SystemC designs. A back-end should focus purely on the behavior of the considered SystemC program. This behavior is fully captured by the SystemC processes under the simulation semantics of the SystemC kernel. Therefore, a front-end should first perform the elaboration phase, i.e. determine the binding of ports and channels. Then it should extract and map the

---

[2]It becomes sufficient if all possible inputs are considered too, e.g. by employing *symbolic simulation*.

design behavior to the IVL. Separating the verification process of SystemC programs into two independent tasks makes both of them more manageable. In the following the structure and key components of the IVL are briefly discussed.

Based on the SystemC simulation semantics described in the previous section, three basic components of the SystemC kernel can be identified: *SC_THREAD*, *sc_event* and *channel update*. These are adopted to be *kernel primitives* of the IVL: *thread*, *event* and *update*, respectively. Associated to them are the following primitive functions in the IVL:

- `suspend` and `resume` to suspend and resume a thread, respectively

- `wait` and `notify` to wait for and notify an event (the notification can be either immediate or delayed depending on the function arguments, similar to the corresponding functions in SystemC)

- `request_update` to request an update to be performed during the update phase

These primitives form the backbone of the kernel. Other SystemC constructs such as *sc_signal*, *sc_mutex*, static sensitivity, etc. can be modeled using this backbone. The behavior of a *thread* or an *update* is defined by a function. Functions which are neither *threads* nor *updates* can also be declared. Every function possesses a body which is a list of statements. A statement is either an assignment, (conditional) goto statements or function call. Every structural control statement (`if-then-else`, `while-do`, `switch-case`, etc.) can be mapped to conditional `goto` statements (this task should also be performed by the front-end). Therefore, the representation of a function body as a list of statements is general and at the same time much more manageable for a back-end. As *data primitives* the IVL supports boolean and integer data types of C++ together with all arithmetic and logic operators. Furthermore, arrays and pointers of primitive types are also supported. For verification purpose, the IVL provides the *assert* and *assume* functions. Symbolic values of integer types and arrays are also supported.

2.2.1. Example

Basically an IVL program consists of a list of declarations. These include functions, threads, global variables and events. The execution of an IVL program starts by evaluating all global variable declarations. Then the (unique) *main* function will be executed. The *start* statement[3] starts the actual simulation. An optional maximum simulation time can be passed as argument. If none or a negative value is passed, the simulation will not be time bounded. The simulation semantics directly correspond to those of SystemC, as described in Section 2.1.1.

Remark. The syntax for the *main* function and threads is slightly different to normal functions, since they neither take arguments nor return a result. Internally all of them are represented as functions though.

In the following an example SystemC and corresponding IVL program are presented. The main purpose of the example is to demonstrate some elements of the IVL. The example appeared in similar form in [Le+13]. It is presented here for convenience. For the sake of clarity, in this example and also in the following high level control structures are used in IVL programs instead of (conditional) gotos. Some slight syntactic adaptions have been performed to make the IVL easier to parse and read.

---

[3]One can think of it as a function too.

```
1  SC_MODULE(Module) {                    1   event e;
2    sc_core::sc_event e;                 2   uint x = ?(uint);
3    uint x, a, b;                        3   uint a = 0;
4                                         4   uint b = 0;
5    SC_CTOR(Module)                      5
6      : x(rand()), a(0), b(0) {          6   thread A {
7      SC_THREAD(A);                      7     if (x % 2)
8      SC_THREAD(B);                      8       a = 1;
9      SC_THREAD(C);                      9     else
10   }                                    10      a = 0;
11                                        11  }
12   void A() {                           12
13   if (x % 2)                           13  thread B {
14     a = 1;                             14    wait e;
15   else                                 15    b = x / 2;
16     a = 0;                             16  }
17   }                                    17
18                                        18  thread C {
19   void B() {                           19    notify e;
20   e.wait();                            20  }
21    b = x / 2;                          21
22   }                                    22  main {
23                                        23    start;
24   void C() {                           24    assert (2 * b + a == x);
25   e.notify();                          25  }
26   }
27 };
```

Listing 2.2: The example program of Listing 2.1
in IVL

```
28
29 int sc_main() {
30   Module m("top");
31   sc_start();
32   assert(2 * m.b + m.a == m.x);
33   return 0;
34 }
```

Listing 2.1: A SystemC example program

SystemC example   Listing 2.1 shows a simple SystemC example. The design has one module and three SC_THREADs A, B and C. Thread A sets variable a to 0, if x is divisible by 2, and to 1 otherwise (line 13-16). Variable x is initialized with a random integer value on line 6 (i.e. it models an input). Thread B waits for the notification of event e and sets $b = x / 2$ subsequently (line 20-21). Thread C performs an immediate notification of event e (line 25). If thread B is not already waiting for it, the notification is lost. After the simulation the value of variable a and b should be $x \% 2$ and $x / 2$, respectively. Thus the assertion $(2 * b + a == x)$ is expected to hold (line 32). Nevertheless, there exist counter-examples, for example the scheduling sequence CAB leads to a violation of the assertion. The reason is that b has not been set correctly due to the lost notification.

IVL example   Listing 2.2 shows the same example in IVL. As can be seen the SystemC module is "unpacked", i.e. variables, functions, and threads of the module are now global declarations. The calls to wait and notify are directly mapped to statements of the same name. Variable x is initialized with a symbolic integer value (line 2) and can have any value in the range of *unsigned int*. The statement start on line 24 starts the simulation.

2.3. Symbolic Execution

In principle symbolic execution [Kin76; CDE08] is similar to normal execution. A program is simulated by executing its statements one after another. The difference is that symbolic execution allows to store and manipulate both symbolic and concrete values. A symbolic value can represent all or a subset of possible concrete value for each state part in the program. Thus it can be used to exhaustively check a single execution path for some or all input data. During execution a path condition pc is managed for every path. This is a Boolean expression that is initialized as $pc = True$. It represents constraints that the symbolic values have to satisfy for the corresponding path, thus effectively selecting the possible values a symbolic expression can evaluate to.

There are basically two different ways to extend a path condition: either by adding an assumption or executing a conditional goto. In order to add an assumption c, which itself is just a boolean expression, to the current path condition pc, e.g. by executing an *assume* statement, the formula $pc \wedge c$ will be checked for satisfiability by e.g. an SMT solver. If it is satisfiable, the path condition is update as $pc := pc \wedge c$. Otherwise the current execution path is considered *unfeasible* and will be terminated.

When a conditional goto with branch condition c is executed in a state s, which represents an execution path, an i.e. SMT solver is used to determine which of the branch condition and its negation is satisfiable with the current path condition. If both are satisfiable, which means both branches are *feasible*, then the execution path s is *forked* into two independent paths. One that will take the goto s_T and one that will not s_F. The path conditions of both paths are updated accordingly as $pc(s_T) := pc(s_T) \wedge c$ and $pc(s_F) := pc(s_F) \wedge \neg c$ respectively.

The symbolic execution effectively creates a tree of execution paths where the path condition represents the constraints under which a specific position in the program will be reached. In order to check whether an assertion of condition c is violated, the formula $pc \wedge \neg c$, where pc is the path condition under which the assertion is reachable, will be checked for satisfiability. The assertion is violated iff the formula is satisfiable.

The combination of symbolic execution with complete exploration of all possible process scheduling sequences enables exhaustive exploration of state spaces. The combined approach is called *symbolic simulation*. It is used by the symbolic simulator *SISSI* [Le+13] to discover assertion violations and other types of errors, such as memory access errors in SystemC IVL programs, or prove that none of them exists.

2.4. State Transition System

A state transition system (STS) is a finite state automaton describing all possible transitions of a system. In principle the definitions with regard to an STS follow the example of [FG05; KGG08; EP10].

Definition 1 (*State Transition System (STS) or State Space*)

A state transition system *(STS)*, or state space, is a five tuple $A = (S, s_0, T, \Delta, s_e)$ where S is a finite set of states; $s_0 \in S$ is the initial state of the system; T denotes all possible transitions; $\Delta \subseteq S \times T \times S$ is the transition relation; $s_e \in S$ is a unique distinguished error state.

Let $A=(S, s_0, T, \Delta, s_e)$ be an STS. If $(s,t,s') \in \Delta$ than s' is the (unique) *successor* of s when executing transition t. It will be denoted as $s \xrightarrow{t} s'$. Sometimes the notation $s \xrightarrow{t}$ will be used to refer to s'. The transition t is said to be *enabled* in s. The set of enabled transitions in a state s is defined as $enabled(s) = \{t \mid (s,t,s') \in \Delta\}$. A state s with $enabled(s) = \emptyset$ is called *deadlock* or *terminal* state. The function $enabled(s)$ will sometimes be abbreviated as $en(s)$.

A (finite) sequence of transitions $w = t_1..t_n \in T^*$ is called a *trace*. A trace is executable from $s \in S$ iff there exists a sequence of states $s_1..s_{n+1}$ such that $s = s_1$ and $t_i \in en(s_i)$ and $s_i \xrightarrow{t_i} s_{i+1}$ for $i \in \{1..n\}$. The notation $s_1 \xrightarrow{w=t_1..t_n} s_{n+1}$ can be used to express the above situation. Thus the notation $s \xrightarrow{w} s'$ can be used for single or multiple transitions. Normally it is clear from the context which definition is used.

A state s' is said to be *reachable* from s iff there exists a trace w such that $s \xrightarrow{w} s'$, which can also be written as $s \xrightarrow{*} s'$ if the actual trace w is irrelevant. A state s is reachable in an STS iff $s_0 \xrightarrow{*} s$. Two transitions t and t' are *co-enabled* if they are both enabled in some reachable state s. Two traces $w_1 = t_1..t_n$ and $w_2 = a_1..a_m$ can be concatenated by the $\cdot$ operation, thus $w_1 \cdot w_2 = t_1..t_n a_1..a_m$. Concatenation can also be used to prepend or append a single transition to a trace.

In the following the term *state space* will be used synonymously to refer to an STS. Sometimes it will even be referred to as (transition) *automaton*. The complete (global) state space is denoted as A_G to distinguish it from the reduced state space A_R that will be defined later during the presentation of state space reduction techniques. Instead of writing $s \in S$ it will often be simply said that s is in A or even $s \in A$, where A is an STS.

As a convention the existence of a (single) distinguished *error state* $s_e \in S$ is assumed. Once an error state is reached, the system will not leave it anymore, thus $\forall t \in T : (s_e, t, s_e) \in \Delta$. The predicate $\perp (s)$ will return True iff s is an error state, which by convention means that $s = s_e$. A transition that violates an assertion during execution will lead to an error state.

Remark. Due to symbolic execution, a single transition can lead to multiple successor states, when a conditional goto is executed where the branch condition and its negation are both satisfiable. For simplicity it will be assumed, and has been in the above description, that each transition has a single unique successor for each state. This is not a real limitation of the theoretical framework, since every non-deterministic automaton can be transformed into a deterministic one. Also this extension can be integrated quite naturally into the state space exploration algorithms that will be presented in this thesis, since all successor states are independent of each other. In practice, multiple successor states that arise due to symbolic execution can be handled quite efficiently. A state space exploration algorithm that explicitly handles multiple successors due to symbolic execution is provided in the Appendix. It will be specifically referred to during the presentation of state space exploration algorithms in Chapter 3.

2.4.1. Modeling IVL Programs as STS

As described in Section 2.2 the behavior of a SystemC program can be represented as IVL program. This section describes how finite IVL programs, i.e. the number of different states is finite and each transition runs for a finite number of steps, can be formally modeled as state transition systems.

A transition moves the system from one state to a subsequent state. In the IVL (and analogously SystemC) basically two different kinds of transitions can be identified, that change the state of the system in the simulation phase: *thread* and *notification* transitions.

Thread transitions change the state by executing a finite sequence of operations of a chosen thread followed by a context switch operation or termination of the same thread. Thus every thread can be separated into a list of transitions. The first transition begins with the first statement of the thread. All other transitions continue from a context switch statement that has interrupted the previous transition. All transitions either end with a context switch or the termination of the thread, which means that all statements of the process have been completely executed. Thus every thread T, that has not been fully executed, has a unique currently active

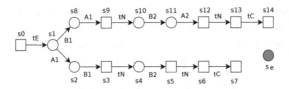

Figure 2.2.: Complete state space for the program in Listing 2.3

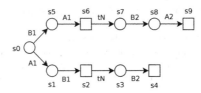

Figure 2.3.: Simplified state space for the program in Listing 2.3

transition in each state s, denoted as $next(s, T)$. This transition can either be enabled or disabled in s. Basically thats the transition that will be executed next when thread T is selected for execution in state s. A transition is enabled in a state s if it is the current active transition of a runnable thread.

For verification purposes, the SystemC IVL supports the *assume* and *assert* statements. Whenever an assertion violation is detected during the execution of a thread, the system will reach the designated error state s_e. The assume statement can be handled similarly, by introducing a designated terminal state, that is reached whenever the assumed condition is unsatisfiable.

A notification transition changes the state of the system by performing the *update, delta notify* and *timed notification* phases as necessary. The notification transition will be denoted as t_N in the following. A state s where $t_N \in en(s)$ is called a *notification* or *notify* state. According to the simulation semantics of SystemC, a notification transition will only be enabled if no thread transitions are runnable. Thus $t_N \in en(s)$ always implies that $en(s) = \{t_N\}$. All transitions between two notification states belong to the same *delta cycle*.

Thread and notification transitions are sufficient to model the simulation phase, which begins with the *start* statement. The execution of statements before and after the *start* statement can be modeled by introducing two additional distinguished transitions t_E and t_C respectively, similarly to the notification phase transition t_N. For the sake of simplicity these transitions will not be further considered in the following. The next section provides an example to illustrate the concepts of this section.

2.4.2. Example

Consider the simple IVL program in Listing 2.3. It consists of two threads A and B. Both of them consist of two transitions, separated by context switches, called A_1, A_2 and B_1, B_2 respec-

```
1  int a = 0;            8    }                     15    notify eA;
2  int b = 0;            9      assert (b == 0);    16  }
3                        10   }                     17
4  thread A {            11                         18  main {
5    if (b > 0) {        12  thread B {            19    start;
6      wait eA;          13    b = 1;               20  }
7      b = 0;            14    wait_time 0;
```

Listing 2.3: Example to demonstrate the correspondence between IVL programs and state transition systems

tively. The transitions can[4] execute the statements at lines:

$$A_1 = \{5,6,9\}$$
$$A_2 = \{7,9\}$$
$$B_1 = \{13,14\}$$
$$B_2 = \{15\}$$

The STS for this program is defined as $A=(S, s_0, T, \Delta, s_e)$ with:

$$T = \{A_1, A_2, B_1, B_2, t_C, t_E, t_N\}$$
$$\Delta = \{(s_0, t_E, s_1), (s_1, A_1, s_2), (s_1, B_1, s_8), (s_2, B_1, s_3), (s_3, t_N, s_4), (s_4, B_2, s_5), (s_5, t_N, s_6),$$
$$(s_6, t_C, s_7), (s_8, A_1, s_9), (s_9, t_N, s_{10}), (s_{10}, B_2, s_{11}), (s_{11}, A_2, s_{12}), (s_{12}, t_N, s_{13}),$$
$$(s_{13}, t_C, s_{14})\}$$
$$S = \{s_0, s_1, s_2, s_3, s_4, s_5, s_6, s_7, s_8, s_9, s_{10}, s_{11}, s_{12}, s_{13}, s_{14}, s_e\}$$

A graphical representation of the STS is shown in Figure 2.2. Circles denote normal states, i.e. states where a transition can be selected, whereas squares denote states where the either one of the designated transitions $\{t_E, t_C, t_N\}$ is explored. In the following the t_E and t_C transitions will not be explicitly considered. Furthermore the last notification transition and unreachable error state will also normally be omitted. A graphical representation incorporating these simplifications is shown in Figure 2.3. Sometimes even the notification transitions in between will be omitted, since they can be unambiguously inferred given the original program. Error states will be shaded, as shown in Figure 2.2.

2.4.3. Remarks

This modeling requires not only that a SystemC program is finite but also that every transition considered in isolation will terminate. For example consider the simple IVL program in Listing 2.5. It would be invalid according to the above definition, since the transition of the thread A will not terminate, thus not reach a successor state. On the other hand a program with finite cyclic state space as shown in Listing 2.4 is valid, since every transition will eventually either finish execution or hit a context switch. While programs that loop without context switches are very interesting and challenging in the context of formal verification of generic programs, they are rather uncommon in the context of SystemC. For this reason such programs will not be further considered, though the methods presented in this thesis can be extended to support them.

---

[4]Whether the transition A_1 takes the branch in Line 5 depends on the state it is executed from.

```
1  thread A {                          1  thread A {
2    while (true) {                    2    while (true) {
3      wait_time 0;                    3    }
4    }                                 4  }
5  }                                   5
6                                      6  main {
7  main {                             7    start;
8    start;                            8  }
9  }
```

Listing 2.4: IVL program that can be modelled as STS.

Listing 2.5: IVL program that cannot be modeled as STS due to non-terminating transitions.

Algorithm 1: Complete Stateful DFS

Input: Initial state

1 $H \leftarrow Set()$
2 $explore(\texttt{initialState})$

3 **procedure** $explore(s)$ **is**
4   **if** $s \notin H$ **then**
5        $H.add(\texttt{s})$
6   **for** $t \in en(s)$ **do**
7           $n \leftarrow succ(s,t)$
8           $explore(\texttt{n})$

2.5. Basic Stateful Model Checking Algorithm

This section presents a basic stateful DFS algorithm that will explore the complete state space. It is shown in Algorithm 1. A set H is managed to store already visited states. The algorithm starts by calling *explore* with the initial state.

The *explore* procedure takes a state s as argument and will recursively explore all reachable states from s. First it checks whether s has already been explored. If not, s is added to the set of visited states H (line Line 5) and all enabled transitions in s are recursively explored one after another.

This basic algorithm performs a complete stateful exploration and thus suffers from the well-known state explosion problem. As has already been mentioned, POR and SSR will be applied in this thesis to alleviate the problem. The former explores only a subset of enabled transitions in Line 6. The latter uses a different equality predicate for the check $s \notin H$ in Line 4. Both reduction techniques can be applied together.

2.6. Partial Order Reduction

Partial Order Reduction (POR) is a widely used and particularly effective technique to combat the state explosion problem that arises in model checking of concurrent systems. The idea is to explore only a subset of the complete state space that is provably sufficient to verify the properties of interest.

It is based on the observation, that concurrent systems allow for the execution of many different transition interleavings, which yield the same resulting state. Thus it is sufficient to explore the execution of these equivalent transition sequences in a single representative order, reducing resource usage required for model checking.

Several similar approaches for POR have been developed. Most of them work by selecting a subset of enabled transitions in each state, resulting in the exploration of a reduced state space. Such sets are called *persistent* [God91; GP93; God96], *stubborn* [Val89; Val91; Val98; VV98] or *ample* [Pel93; Cla+99]. Optionally *sleep* sets can be additionally used to select smaller subsets of enabled transitions in each state, thus further reducing the explored state space [GW93; God96]. The persistent set framework as presented in [God96] will be used in this thesis.

The reduction is achieved on the fly during state space exploration to reduce the resource usage. Constructing the complete state space in advance and then verify properties of interest on the reduced part of the state space would defy the whole purpose of the reduction.

Two different approaches are commonly used to compute persistent sets for each state. Static- and Dynamic Partial Order Reduction, denoted as SPOR and DPOR respectively. The former is a classical approach, e.g. from [God96], that computes persistent sets whenever a state is entered, based on some statically precomputed transition interference relations. The latter is a more recent approach, first introduced in [FG05], that explores an arbitrary path until completion and dynamically tracks runtime informations to compute dependencies. These are used to detect and explore alternate relevant interleavings of transitions. Ultimately, DPOR will also have explored a persistent set from each state.

The advantage of DPOR is that the information at runtime are precise, this allows to compute dependencies more accurately without too much over-approximation, resulting in a better reduction. Especially when arrays and pointers are involved [5]. A disadvantage compared to the static method is the additional runtime overhead to track runtime informations and compute dependencies. This is necessary, since the actual statements executed by a transition depend on the state where the transition is executed from. Also a DPOR algorithm does not necessarily yield a better reduction than a SPOR algorithm. The achieved reduction depends on the actual implementation.

Another notable advantage of SPOR compared to DPOR is that integration of complementary techniques like stateful model checking, state subsumption matching and parallel execution among others is easier with SPOR [Bok+11; YWY06; Yan+08; Yan+07; Sim+13]. The DPOR reduction requires non-trivial extensions in each case to preserve the soundness of the combined approach. To guarantee the soundness, often conservative assumptions are used thus reducing the benefit of the higher precision of DPOR compared to SPOR. This issue will be further discussed when presenting DPOR in Chapter 4 and State Subsumption Reduction (SSR) in Chapter 5.

In the following some standard definitions with respect to POR are introduced. They include the definition of reduced state spaces, transition interference, trace equivalence and persistent sets. Then Section 2.6.2 presents algorithms to compute static persistent sets based on (statically) precomputed transition interference relations. Finally, a stateless DPOR in the context of SystemC is introduced.

---

[5]Even though the informations at runtime are precisely available, the computed dependencies does not necessarily have to be precise, e.g. it is inherently complex to precisely decide whether two array accesses $a[i]$ and $a[k]$ must overlap if i and k are symbolic integer values.

2.6.1. Definitions

Reduced State Space

Partial order reduction explores only a subset of the complete state space that is sufficient to verify all properties of interest. Most approaches work by selecting only a subset of enabled transitions in each state. This can be formalized by means of a reduction function r, e.g. as shown in [EP10; BLL06].

Definition 2 (*Reduction Function*)

> A reduction function r for an STS $A=(S, s_0, T, \Delta, s_e)$ is a mapping from S to 2^T such that $r(s) \subseteq en(s)$.

When all enabled transitions are in $r(s)$ for some state s, so $en(s) = r(s)$, then no reduction is provided for state s. The state s is said to be *fully expanded*. A transition t is *postponed* in s if $t \in en(s) \setminus r(s)$.

A reduced state space A_R can be constructed by applying a reduction function r to each reachable state s of another state space A. Doing so will yield a subset of enabled transitions $r(s) \subseteq en(s)$ for each state s. Only transitions $t \in r(s)$ will be explored.

Definition 3 (*Reduced STS*)

> Let $A=(S, s_0, T, \Delta, s_e)$ be an STS. A reduced STS $A_R=(S_R, s_{0R}, T_R, \Delta_R, s_{eR})$ can be defined by means of a reduction function r as:
>
> - $s_{0R} = s_0$ and $T_R = T$ and $s_e = s_{eR}$
>
> - $s \in S_R$ iff there exists an execution sequence $s_0...s_n$ such that $s = s_n$ and $t_i \in r(s_j)$ for $j \in \{1..n\}$
>
> - $(s, t, s') \in \Delta_R$ iff $s \in S_R$ and $t \in r(s)$ and $(s, t, s') \in \Delta$

The reduced state space contains a subset of states and transitions of the original state space, thus $S_R \subseteq S$ and $\Delta_R \subseteq \Delta$. Normally the reduced state space A_R is computed from the complete state space A_G.

Hereafter the notations $s \xrightarrow{t}_R s'$, $s \xrightarrow{*}_R s'$ and $s \xrightarrow{t}_R$ will sometimes be used to explicitly refer to the reduced state space. Normally they will be omitted though, since it is clear from the context to which state space they refer.

Selective Search

This section presents a simple algorithm that will explore all states of a reduced state space A_R with the reduction function r. Such a selective search forms the basis of all following algorithms and might be more intuitive to use than the declarative definition of the reduced state space. The algorithm is shown in Algorithm 2. It employs a stateful depth first search to visit all reachable states[6]. Already visited states are stored in the (hash-)set H to avoid re-exploration. Compared to the basic stateful model checking algorithm shown in Algorithm 1, successor states are generated by means of the reduction function r, which selects only a subset

---

[6]Using a depth first search is just a design decision, because it allows for a memory efficient state space exploration. Any search method could be used here. It is only relevant that the complete reduced state space is explored.

Algorithm 2: Selective Stateful DFS

Input: Initial state

```
1  H ← Set()
2  explore(initialState)

3  procedure explore(s) is
4      if s ∉ H then
5          H ← H ∪ s
6          for t ∈ r(s) do
7              n ← succ(s, t)
8              explore(n)
```

of enabled transitions for each reached state. Hence it is called a *selective search* [God96; EP10].

Transition Interference

The idea of (in-)dependence of transitions is a common point of many POR algorithms. Intuitively, two transitions t_1 and t_2 are independent if they cannot disable each other and if they commute in any state of the system [FG05; KGG08; EP10].

Definition 4 (*Transition (In-)Dependence*)

> *A transition independence relation is a irreflexive, symmetric relation $I \subseteq T \times T$ satisfying the following two conditions for each state $s \in S$ in the complete state space A_G and transitions $(t_1, t_2) \in I$:*
>
> 1. *Enabledness: If $t_1, t_2 \in en(s)$ and $s \xrightarrow{t_1} s'$ then $t_2 \in en(s')$*
>
> 2. *Commutativity: If $t_1, t_2 \in en(s)$ and $s \xrightarrow{t_1 t_2} s'$ then $s \xrightarrow{t_2 t_1} s'$*
>
> *Given an independence relation I, a dependency relation $\text{dep} \subseteq T \times T$ can be obtained as $\text{dep} = T \times T \setminus I$.*

By definition transitions that are not co-enabled are always independent, since no state $s \in S$ exists such that $t_1, t_2 \in en(s)$. All transitions that cannot be proven independent are assumed to be dependent.

Another useful transition interference relation is the *can-enabling* relation, denoted as *ce*, which appeared in similar form in [Bok+11]. It is based on the concept of *necessary enabling transitions* (NES) [God96]. A transition t_1 *can enable* another transition t_2, if in at least one state where t_2 is disabled, executing t_1 results in a state where t_2 is enabled.

Definition 5 (*Transition* can-enable)

> *A transition can-enable relation is a irreflexive relation $ce \subseteq T \times T$ satisfying*
>
> $$ce \supseteq \{(t_1, t_2) \in T \times T \mid \exists s, s' \in S : t_1 \in en(s) \text{ and } t_2 \notin en(s) \text{ and } s \xrightarrow{t_1} s' \text{ and } t_2 \in en(s')\}$$
>
> *for all states S in A_G.*

These transition interference relations can either be computed before the exploration of the state space on the basis of a static analysis of the program or can be dynamically inferred at runtime. The former approach is taken by a SPOR method, whereas the latter approach is employed by a DPOR method. It is always safe to over-approximate the dependency and can-enable relation, i.e. including pairs of non-interfering transitions. Though more precise definitions of transition interference relations are a crucial part of every POR method, as they are used as basis to compute the subset of enabled transitions that shall be explored for each state. Thus more precise relations yield better reductions and can result in the exploration of a smaller state space.

IVL specific transition interference   The above definitions are generic, as no specific semantic has been used. This section instantiates them to the concrete semantics of the IVL.

Two transitions t_1 and t_2 are dependent if they are co-enabled and either (a) access the same memory location where at least one non-commutative write access is involved (b) or one of them can immediately notify the other (c) or one of them can suspend the other. All of these conditions lead to the result that t_1 and t_2 do not commute. Condition (c) additionally states that one transition disables the other. For simplicity it simply is assumed that all transitions are co-enabled. Condition (a) considers some bitvector operations like addition, multiplication and logic operations to be commutative.

A transition t_1 can enable another transition t_2 if t_1 and t_2 are co-enabled in the same delta cycle and t_2 waits on an event e and t immediately notifies e. Again for simplicity it is simply assumed that all transitions can be co-enabled in the same delta cycle. Doing so is always safe, since it can only lead to over-approximation of the transition dependency relation.

Delta- and timed notifications can also enable transitions, but they are not considered for the computation of the dependency and can-enable relation. The effects of such notifications are postponed until the next delta cycle. The simulation can only proceed from one delta cycle to the next, if a state is reached, where no enabled (runnable) transition is available. Consequently all these transition sequences are equivalent. For this reason only notifications whose effects happen during the same delta cycle are considered in the *can-enable* relation ce. It turns out that only immediate notifications fall into this category.

Trace Equivalence and Partial Orders

This section will formalize the notion of trace equivalence and relate it to partial orders. The description follows the notion presented in [God96]. First the definitions will be presented, then some examples will be shown to illustrate the concepts. According to [Maz87; God96] transition sequences (traces) can be grouped into equivalence classes by means of an (in-)dependence relation.

Definition 6 (*Trace Equivalence*)

> Two sequences of transitions w_1 and w_2 are equivalent, denoted as $w_1 \equiv w_2$, if they can be obtained from each other by successively permuting adjacent independent transitions.

The set of equivalent transition sequences for w will be denoted as $[w] = \{w' \in T^* \mid w' \equiv w\}$. It is called the trace (equivalence) class of w. For example let $w_1 = t_1 t_2 t_3$ and $w_2 = t_2 t_1 t_3$ and let t_1 be independent with t_2, then $w_1 \equiv w_2$ holds. Since $\equiv$ is an equivalence relation by definition, $w_1 \in [w_2]$, $w_2 \in [w_1]$ and $[w_1] = [w_2]$ hold too. By successively permuting adjacent independent transitions in w, one can obtain all transition sequences in the trace class $[w]$. Thus a trace class is fully characterized by one of its members and a (in-)dependence relation between transitions. By definition, all transition sequences in a given trace class contain the same number of transitions (since they can be obtained from each other by swapping adjacent transitions).

Moreover every one of them will result in the same state, when executed from the same state, as the following theorem establishes [God96].

Theorem 2.6.1

Let s be a state in A_G. If $s \xrightarrow{w_1} s_1$ and $s \xrightarrow{w_2} s_2$ in A_G, and if $[w_1] = [w_2]$, then $s_1 = s_2$.

The reader is referred to [God96] for a proof of this theorem. In the following, the notation $[w]_s$ will often be used to denote the trace equivalence class of w when executed from state s.

Remark. The term *trace* refers in this thesis to a sequence of transitions. Sometimes it will also be called a path. In [Maz87; God96] the term *trace* refers to the complete equivalence class $[w]$, thus it includes (potentially) multiple sequences of transitions. In this thesis the equivalence class $[w]$ is referred to as either *trace class* or more commonly simply as *trace equivalence class*.

Definition 7 (*Partial Order*)

A relation $R \subseteq A \times A$ on a set A is a partial order iff R is reflexive, antisymmetric and transitive. A partial order R is also a total order if for all $a_1, a_2 \in A$ either $(a_1, a_2) \in R$ or $(a_2, a_1) \in R$.

A partial order can be extended to a total order by a linearization.

Definition 8

A linearization of a partial order $R \subseteq A \times A$ is a total order $R' \subseteq A \times A$ such that $R \subseteq R'$.

The intersection of all linearizations of a partial order results again in the same partial order. There is a correspondence between trace equivalence classes and partial orders of transition occurrences. The set of transition sequences in the trace class is the set of all linearizations of the partial order of transition occurrences [God96]. Partial orders and thus trace equivalence classes can be naturally represented by a *happens before* relation, e.g. as shown in [KGG08; FG05]:

Definition 9 (*Happens Before*)

Let $w = t_1..t_n$ be a trace in A_G. Let dep be a dependency relation between transitions. A happens before relation $\trianglelefteq_w$ is the smallest binary relation on T such that:

1. if $i \le j$ and $(t_i, t_j) \in$ dep then $i \trianglelefteq_w j$

2. $\trianglelefteq_w$ is transitively closed

Example Consider the set $T = \{t_1, t_2, t_3, t_4\}$ of transitions. Let the independence relation I be defined as the symmetric closure of $I_0 = \{(t_2, t_3), (t_2, t_4)\}$. Thus all other transition pairs are dependent. Let s be a state in A_G and $w = t_1 t_2 t_3 t_1 t_2 t_4 t_1$ a trace in A_G from s. The trace equivalence class with respect to I is defined as:

$$[w]_s = \{t_1 t_2 t_3 t_1 t_2 t_4 t_1,\ t_1 t_2 t_3 t_1 t_4 t_2 t_1,\ t_1 t_3 t_2 t_1 t_2 t_4 t_1,\ t_1 t_3 t_2 t_1 t_4 t_2 t_1\}$$

All traces in $[w]_s$ lead to the same result state from s. The trace w contains seven transition occurrences of four different transitions. The occurrences can be explicitly named as $w_a = a_1..a_7$. Thus e.g. a_1 corresponds to the first occurrence of t_1, and a_4 to the second occurrence of

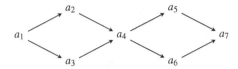

Figure 2.4.: Partial order of transition occurrences for the trace w and independence relation I

t_1 and so on. A partial order R that captures the (in-)dependencies of the transition occurrences is the transitive, reflexive closure of P which is defined as:

$$P = \{(a_1,a_2),(a_1,a_3),(a_2,a_4),(a_3,a_4),(a_4,a_5),(a_4,a_6),(a_5,a_7),(a_6,a_7)\}$$

Basically P encodes the dependencies between the transition occurrences. A graphical representation, that omits reflexive and transitive edges, is shown in Figure 2.4. An edge denotes that the two transition occurrences are dependent. The partial order also has four different linearizations. They correspond exactly to the trace equivalence class $[w]_s$, when replacing transition occurrences with the actual transitions. Similarly the happens before relation of transition occurrences $\trianglelefteq_w$ corresponds with the partial order R obtained from P. It is defined as $\trianglelefteq_w = \{(i,j) \mid (a_i,a_j) \in R\}$, thus e.g. $1 \trianglelefteq_w 2$, $1 \trianglelefteq_w 7$, $3 \trianglelefteq_w 5$, and so on.

Persistent Set

Intuitively a set of transitions T is persistent in a state s of A_G, if starting from s and executing only transitions not in T, will have no dependency to any transition in T. So whatever one does from s while staying outside of T will not influence any transition in T. Consequently at least one transition $t \in T$ has to be executed from s or any state reachable from s to produce a dependency with any transition in the persistent set T. Persistent sets are formally defined as [God96]:

Definition 10 (Persistent Set)

A set T of transitions enabled in a state s is persistent in s iff, for all non empty sequences of transitions

$$s = s_1 \xrightarrow{t_1} s_2 \xrightarrow{t_2} \dots s_n \xrightarrow{t_n} s_{n+1}$$

from s in A_G and including only transitions $t_i \notin T$, for $i \in \{1,...,n\}$, t_n is independent in s_n with all the transitions in T.

A persistent set in s is called *trivial*, if it contains all transitions enabled in s.

2.6.2. Computing Static Persistent Sets

This section presents two different algorithms that appeared in [God96] to compute persistent sets using statically precomputed transition interference informations (dependency and can-enable relations).

Conflicting Transitions Algorithm

This section presents an algorithm, based on the conflicting transitions algorithm of [God96], to compute simple non trivial persistent sets. It only requires a valid dependency relation dep between transitions.

Algorithm 3: Conflicting Transitions Algorithm

Input: State s and any enabled transition t_s (any $t_s \in en(s)$)
Output: Persistent set in s

1    $working \leftarrow t_s$
2    $done \leftarrow \{\}$
3   **while** $|working| > 0$ **do**
4   t $\leftarrow$ working.pop
5   **if** $t \notin done$ **then**
6         $done.add(\text{t})$
7   **if** $t \in en(s)$ **then**
8           $working.update(\text{dep[t]})$
9   **else**
10   **return** $en(s)$
11 **return** $done$

The algorithm computes a set of transitions T that can be inductively defined. Initially $T = \{t_s\}$ where t_s is an arbitrary enabled transition in state s. Then $\forall t \in T$:

- $t \in en(s) \implies dep[t] \subseteq T$

- $t \notin en(s) \implies en(s) \subseteq T$

The algorithm then returns $en(s) \cap T$, which is a persistent set in s. A concrete implementation of this description is available in Algorithm 3.

Whenever a dependency to a disabled transition t is encountered, the algorithm aborts and just returns a trivial persistent set in s (all enabled transitions in s). The reason is, that the algorithm does not know, which transition may enable t. Thus it simply assumes that any transition could do so. A simple example that demonstrates why it is necessary to include transitions which can enable currently disabled transitions into the persistent set is given in Example 1. The algorithm in the next section additionally makes use of the ce relation to compute non trivial persistent sets even if a disabled transition is encountered.

Example 1. This example demonstrates why it is necessary to include transitions, which can enable currently disabled transitions, into the persistent set. Not doing so can miss assertion violations as will be shown in the following. Consider the program in Listing 2.6 It consists of three threads A,B and D. Both threads A and B consists of a single transition which will also simply be denoted as A and B. Thread D consists of two transitions denoted as D_0 and D_1. The dependency relation is given as dep $= \{(A,D_1),(B,D_0)\}$ and the can-enable relation is ce $= \{(B,D_1)\}$. Now assuming that the path $s_0 \xrightarrow{A} s_1 \xrightarrow{B} s_2 \xrightarrow{D_0} s_3$ is explored first, then the enabled transitions en and persistent sets ps for the visited states would be:

$en(s_0) = \{A,B,D_0\}$ $ps(s_0) = \{A\}$
$en(s_1) = \{B,D_0\}$ $ps(s_1) = \{B,D_0\}$
$en(s_2) = \{D_0\}$ $ps(s_2) = \{D_0\}$
$en(s_3) = \{\}$ $ps(s_3) = \{\}$

So only the additional transition sequence A, C_0, B, C_1, starting from the initial state s_0, would be explored. And the transition sequence C_0, B, C_1, A, which leads to an error state from s_0, would be missed.

```
1  int x = 0;        5    assert (x ==      8  thread B {      12  thread D {
2  event e;               0);              9    notify e;     13    wait e;
3                    6  }                 10  }               14    x = 1;
4  thread A {        7                    11                  15  }
```

Listing 2.6: Example program

Stubborn Set Algorithm

This section presents a stubborn set algorithm to compute persistent sets. It works similarly to the above conflicting transitions algorithm by maintaining an initially empty result set and adding dependent transitions to it but it can still return a non trivial persistent set when a disabled transition is introduced into the result set. Consequently this algorithm does always yield persistent sets that are smaller or equal to the persistent sets computed by the above algorithm.

Definition 11 (*Necessary Enabling Set*)

Let t be a transition that is disabled in state s. Let s' be any reachable state from s in A_G, such that t is enabled in s'. A set of transitions T is necessary enabling for t in s, denoted as $nes(s, t)$, iff for all traces w that lead to the state s' from s, w contains at least one transition of T.

So basically a disabled transition t in a state s cannot be executed before at least one of $nes(s, t)$ is executed. With this definition in place a stubborn set can be defined as:

Definition 12 (*Stubborn Set*)

A set T of transitions is a stubborn set in a state s if T contains at least one enabled transition and for all transitions $t \in T$ the following conditions hold:

- If t is disabled in s, then all transitions of one necessary enabling set $nes(s, t)$ are also in T ($t \notin en(s) \implies \exists nes(s, t) : nes(s, t) \subseteq T$).

- If t is enabled is s, then all transitions that directly dependent on t are also in T ($t \in en(s) \implies \{d \mid (t, d) \in dep\} \subseteq T$).

According to [God96] this definition resembles strong stubborn sets as defined in [Val89]. And every stubborn set conforming Definition 12 is also a persistent set.

A concrete generic algorithm that computes stubborn sets according to Definition 12 is shown in Algorithm 4. It is generic in the sense that it only depends on a valid dependency relation dep between transitions and a way to compute necessary enabling transitions. The function nes can be defined based on a valid can-enable relation ce, which has been already defined as: $(t_1, t_2) \in ce$ iff t_1 can enable t_2. Two different implementations nes_1 and nes_2 are given in the following. They receive the current state s and a transition t that is disabled in s as arguments. The first implementation simply returns all transition that can directly enable t. Clearly one of them must be executed in any trace starting from s that leads to the execution of t.

$$nes_1(s, t) = \overleftarrow{ce}[t] = \{(a, b) \mid (b, a) \in ce\}[t] = \{b \mid (b, a) \in ce\}$$

The second implementation exploits the fact that for all states s only a single transition is active for each thread, denoted as $next(s, thread)$. This transition can either be enabled or disabled in

Algorithm 4: Stubborn Set Algorithm Template

Input: State s and any enabled transition t_s (any $t_s \in en(s)$)
Output: Stubborn set satisfying Definition 12

```
1  working ← t_s
2  stub ← {}
3  while |working| > 0 do
4  │   t ← working.pop
5  │   if t ∉ stub then
6  │   │   stub.add(t)
7  │   │   if t ∈ en(s) then
8  │   │   │   working.update(dep[t])
9  │   │   else
10 │   │   │   working.update(nes(s, t))
11 return stub ∩ en(s)
```

state s. It will be the next transition executed for the thread either in s or any of its successor states. Let $t_n = next(s, thread(t))$ be the currently active transition for the thread that contains the transition t. So if $t \neq t_n$, then t_n must be executed before t in every trace starting from s. Consequently $\{t_n\}$ is a necessary enabling set for t in s in that case. Else $t = t_n$, so t is the transition that would be executed next on its containing thread, all transitions are returned that can enable t.

$$
\text{nes}_2(s,t) = \begin{cases} \{t_n\} & \text{if } t \neq t_n \\ \text{nes}_1(s,t) & \text{else} \end{cases}
$$

The stubborn set algorithm using the first implementation will be called STUB$_1$ or simply STUB. When the second implementation is used, it will be called STUB$_2$.

Discussion

Two different algorithms have been presented, the conflicting transitions algorithm and the stubborn set algorithm. Both of them return persistent sets according to Definition 10. Both algorithms take a state s and a transition t that is enabled in s as argument. The stubborn set algorithm will always yield smaller or equal persistent sets compared to the conflicting transitions algorithm.

The stubborn set algorithm is available in two configuration that differ in the way necessary enabling transitions are computed. The first one uses the definition nes$_1$, the second one uses nes$_2$. They will be referred to as $stub_1$ and $stub_2$ respectively. Neither of them is always better than the other one, in the sense that it will always return a smaller persistent set for the same state s and initial transition t. For example consider the example program in Listing 2.7. It consists of four threads A, B, C and D. The threads A, B and C consist only of a single transition denoted as A_0, B_0 and C_0 respectively. The thread D consists of three transitions D_0, D_1 and D_2 separated by the context switches in the lines $\{21, 22\}$. Let s_0 be the initial state, thus $en(s_0) = \{A_0, B_0, C_0, D_0\}$. The dependency and can-enable relations are defined as:

$$
\text{dep} = \{(A_0, D_2), (B_0, D_1), (C_0, D_0)\}
$$
$$
\text{ce} = \{(B_0, D_2), (C_0, D_1)\}
$$

```
1  /*                    6  event e2;          14  }                    22  wait e2;
2  Unsafe program,       7                      15                       23  assert x == 1;
   D -> C -> D           8  thread A {          16  thread C {           24  }
   -> B -> D             9  x = 1;              17  notify e1;           25
   will fail.           10  }                   18  }                    26  main {
3  */                   11                      19                       27  start;
4  int x = 0;           12  thread B {          20  thread D {           28  }
5  event e1;            13  notify e2;          21  wait e1;
```

Listing 2.7: Example program for comparing the stubborn set algorithms

Choosing the enabled transition A_0 in s_0, the first algorithm will yield $stub_1(s_0,A_0) = en(s_0)$ whereas the second algorithm will yield a smaller persistent set $stub_2(s_0,A_0) = \{A_0,C_0,D_0\}$. On the other hand let s_1 be the state reached by executing B_0 from s_0, thus $en(s_1) = \{A_0,C_0,D_0\}$. Choosing the enabled transition A_0 in s_1, the first algorithm will yield $stub_1(s_1,A_0) = \{A_0\}$ but the second algorithm will yield a larger persistent set in this case namely $stub_2(s_1,A_0) = \{A_0,C_0,D_0\}$. So it might be useful to compute both persistent sets $stub_1(s,t)$ and $stub_2(s,t)$ and choose the smaller one of them.

All of the above algorithms compute a persistent set by starting with a transition $t \in en(s)$. Thus the resulting persistent set for each state depends on the initially chosen transition t. Consequently one might consider to compute all possible persistent sets in a state s and choose the smallest one of them. This results in locally optimal persistent sets with respect to the chosen algorithm but not necessarily in globally optimal decisions. Always choosing the smallest persistent set in each reached state does not necessarily lead to the exploration of the smallest possible number of transitions and states, thus it only is a heuristic. Therefore it might even be useful to select a locally non minimal persistent set to achieve a globally maximum reduction [GP93; God96]. It has been shown in [Pel93] that calculating an ample set, which is in principle similar to a persistent set, for each state that leads to a minimal reduced state graph A_R is NP-Hard. Thus the problem of selecting an *optimal* persistent set in each state seems inherently complex.

2.6.3. Dynamic Partial Order Reduction

Dynamic Partial Order Reduction (DPOR) has been introduced in [FG05]. It works by exploring an arbitrary sequence of transitions until completion and dynamically tracks relevant runtime informations to compute dependencies. These are used to detect and explore alternate relevant interleavings of transitions. It supports stateless exploration of acyclic state spaces.

Definitions

Every transition t executed from a state s can be associated with a set of *visible effects*. Visible effects are informations that are relevant to decide whether two transitions are dependent. They contain for example identifiers of variables (or memory locations) which may be shared with other threads and have been accessed, together with the access mode (e.g. read/write). They also contain thread/transition identifiers which have been disabled and events that have been immediately notified. Basically enough information is tracked during the execution of a transition such that given two effect sets e_1 and e_2, it can be decided whether e_1 and e_2 are dependent, according to Definition 4 of transition dependencies. Basically enough information is tracked during the execution of a transition such that given two effect sets e_1 and e_2, the IVL specific transition dependency relation, given in Section 8, can be (conservatively) decided between them. Given a set of visible effects e, the corresponding transition and its thread can be obtained by $transition(e)$ and $thread(e)$ respectively.

Each state s will be associated with a *working* and *done* set. The former contains transitions that needs to be explored from s. The latter contains transitions which have already been explored.

A path/trace is a sequence of transitions. Let $s_1 \xrightarrow{t_1} s_2 \xrightarrow{t_2} ...s_n \xrightarrow{t_n} s_{n+1}$ be a path P. Then the following auxiliary functions are defined:

- $last(P) = s_{n+1}$

- $t \in P$ iff $t = t_i$ with $i \in \{1, n+1..\}$

The thread of a transition t_i can be obtained by $thread(t_i)$. The same transition can occur multiple times in a trace, e.g. t_i can be equal to t_j with $i \neq j$. Though the visible effects naturally can be different for both of them, since s_i does not have to be equal to s_j.

For all states s only a single transition t is active for each thread T. It means that t will be executed next when thread T is selected in s. The function $next(s, T)$ returns the active transition for the T from state s. The active transition for a thread T in state s can either be enabled or disabled. It will not change until the thread T is executed in s or any of its successor states.

A variant of the *happens before* relation introduced in Definition 9 in Section 2.6.1 can be used in a DPOR algorithm to infer dependencies more accurately. Given a thread T and a trace P (as above) it is defined as $i \trianglelefteq_P T$ if

1. $thread(t_i) = T$ or

2. $\exists k \in \{i+1..n\}$ such that $i \trianglelefteq_P k$ and $k \trianglelefteq_P T$

Intuitively, if $i \trianglelefteq_P T$, then the next transition of thread T from the state $last(P)$ is not the next transition of thread T in the state right before transition t_i in either this transition sequence or in any equivalent sequence obtained by swapping adjacent independent transitions [FG05].

Basic Stateless Algorithm

This section presents a basic stateless DPOR algorithm, that can detect all deadlocks and assertion violations of a finite acyclic state space [FG05]. It is based on the SystemC specific adaptions presented in [KGG08]. According to [KGG08] the adapted algorithm preserves all deadlocks and assertion violations.

This section is organized as follows: First a short abstract description of the algorithm is provided. Then the full algorithm will be presented together with a more detailed description. Finally some variations of the algorithm are discussed and an example state space exploration based on this algorithm is presented.

Algorithm Summary The DPOR state space exploration algorithm is based on a DFS search, thus it alternates between transition exploration and backtracking. The algorithm starts by exploring an arbitrary path $s_0 = s_1 \xrightarrow{t_1} s_2 \xrightarrow{t_2} ...s_n \xrightarrow{t_n} s_{n+1}$ (transition sequence) from the initial state s_0 until a terminal state[7] is reached. The visible effects of every executed transition are tracked. Whenever a state is backtracked from the search path, the effects of the last transition t will be compared with all the effects of all relevant previous transitions t_i. If a dependency is detected, backtracking points are added to the state s_i, from which t_i has been explored. Basically it means that s_i is associated with additional transitions $s_i.working$ that need to be explored from

---

[7]A state where no transition is enabled.

Algorithm 5: Stateless DPOR DFS

Input: Initial state
Output: Stateless DPOR DFS state space exploration

1 $P \leftarrow Trace(initialState)$
2 $explore()$

3 **procedure** $explore()$ **is**
4 $\quad$ s $\leftarrow$ last(P)
5 $\quad$ **if** $enabled(s) \neq \emptyset$ **then**
6 $\quad\quad$ s.done $\leftarrow \{\}$
7 $\quad\quad$ s.working $\leftarrow \{selectAny(enabled(s))\}$
8 $\quad\quad$ **while** $|s.working| > 0$ **do**
9 $\quad\quad\quad$ $exploreTransitions($s, s.working$)$

10 **procedure** $exploreTransitions(s, T)$ **is**
11 $\quad$ **for** $t \in T$ **do**
12 $\quad\quad$ s.working $\leftarrow$ s.working $\setminus t$
13 $\quad\quad$ s.done $\leftarrow$ s.done $\cup t$
14 $\quad\quad$ n $\leftarrow$ succ(s, t)
15 $\quad\quad$ e $\leftarrow$ visibleEffects(s, t)
16 $\quad\quad$ P.push(e, n)
17 $\quad\quad$ explore()
18 $\quad\quad$ P.pop()
19 $\quad\quad$ normalBacktrackAnalysis(s, e)

20 **procedure** $normalBacktrackAnalysis(s, e)$ **is**
21 $\quad$ addBacktrackPoints(s, e.disables)
22 $\quad$ commonBacktrackAnalysis(e)

23 **procedure** $commonBacktrackAnalysis(e)$ **is**
24 $\quad$ **for** $s_p, e_p \in$
$\quad\quad$ $transitionsOfLastDeltaCycle(trace)$ **do**
25 $\quad\quad$ **if** $areDependent(e_p, e)$ **then**
26 $\quad\quad\quad$ E $\leftarrow \{thread(t) \mid t \in enabled(s_p)\}$
27 $\quad\quad\quad$ **if** $thread(e) \in E$ **then**
28 $\quad\quad\quad\quad$ T $\leftarrow \{next(s_p, thread(e))\}$
29 $\quad\quad\quad$ **else**
30 $\quad\quad\quad\quad$ T $\leftarrow enabled(s_p)$
31 $\quad\quad\quad$ addBacktrackPoints(s_p, T)

32 **procedure** $addBacktrackPoints(s, T)$ **is**
33 $\quad$ s.working $\leftarrow T \setminus$ s.done

---

s_i. Backtracking stops once a state s is reached which contains unexplored backtracking points. An arbitrary transition $s.working$ is chosen and explored until a terminal state is reached. The search continues until the initial state is backtracked (or an error is detected), which means the complete relevant state space has been explored.

Algorithm Description The algorithm is available in Algorithm 5. It is based on a recursive DFS to explore all reachable states. It does not store already visited states, hence the exploration is called *stateless*, so the same state and all of its successors can be explored multiple times when reached from different paths.

The algorithm manages a global search path P. First P is initialized to contain the initial state (Line 1). Then the *explore* function is called. It returns once all states in P have been fully explored and finally backtracked. Thus it can be recursively used to explore the complete state space.

The *explore* function examines the last state s of P (Line 4). If s is a terminal state, then the function will do nothing (so it returns to the caller which will backtrack the last transition). Else s will be associated with an empty *done* and a *working* set that contains an arbitrary enabled transition. Such a transition must exists, since s is not a terminal state. The *done* and *working* sets are used to track the currently explored and those transitions that still need to be explored respectively for each state on the current search path P.

While the set $s.working$ is not empty, all transitions in $s.working$ will be explored, by the *exploreTransitions* function. It basically recursively invokes the *explore* function and than applies the backtrack analysis for each transition in $s.working$. The *exploreTransitions* function returns when all transitions in *s.working*, and their successors, by further recursive calls, have been explored and backtracked. The dependency analysis that is called for each backtracked transitions might have computed backtrack points for the current state s, so *s.working* can again be non empty. Eventually the while loop will terminate, since only a finite number of transitions is enabled in each state s and every recursive call to *exploreTransitions* explores at least one of them. Once they are all explored, so *s.done* $=$ en(s), no more transitions can be added to *s.working*.

The *exploreTransitions* function takes two arguments, a state s and a set of transitions T, which for this algorithm is always *s.working*. First t is added to the *s.done* set and removed from the *s.working* set. This avoids re-exploration of t from s. Then the successor state n of s when executing t, together with the visible effects e observed during the execution of t, are obtained (lines $\{14, 15\}$). The search path P is extended accordingly with the effects e of t and successor state n. The recursive call to *explore* continues with the last state of P, which has been setup to be n. Once the *explore* function returns, all paths from the next state n have been explored and backtracked. Finally the transition t is backtracked from s. This includes two actions. First the extensions to the search path P before the recursive call to *explore* are reversed in Line 18. So basically the search path is restored to the state it had before the push in line Line 16. It allows to explore the next transition in T in the next loop iteration. And second the backtrack dependency analysis is applied for s and e, by calling the *normalBacktrackAnalysis* function. Basically the effects e from s are compared with all relevant previous effects currently on the search path, which are the effects of those transitions leading to the state s from the initial state s_0.

Remark. The backtrack analysis is triggered after the recursive explore call in line Line 17. This is just a design decision. It could also happen before the explore call. Doing so would immediately compute backtrack points once a transition is explored. The current implementation on the other hand delays the computation of backtrack points until the transition is backtracked from the search path. Doing so seems more reasonable, since a path can be fully explored until a terminal state without computing backtrack points. This allows to detect error states slightly faster.

The function *normalBacktrackAnalysis* is called for backtracking a transition t, which is represented by its visible effects e, from state s. All transitions that have been disabled during the execution of t are added as backtrack points T to s. That means all transition in T that have not yet been explored from s (are not in *s.done*) are added to *s.working*. Then the effects e of t from s are compared with all relevant effects e_p of all transitions t_p and their originating states s_p currently in the search path. All transitions t_p which have been executed in the same delta cycle as t are relevant. They are obtained by using the function *transitionsOfLastDeltaCycle* in Line 24. It returns a possibly empty set of pairs $(s_p = s_i, e_p = e_i)$ for $i \in \{1..n\}$ such that all s_i are states where the simulation is in the evaluation phase and the immediate predecessor s_{i-1} is a state where the simulation is in the notification phase [8]. Whenever a dependency between two effect sets e and e_p is detected, as defined in Section 8, some backtrack points are added. If *thread*(e) is enabled in s_p than a single backtrack point is added, namely the next transition of *thread*(e) that will be executed in s_p. Else the dynamic analysis was unable to infer a non trivial persistent set for s_p and all enabled transitions in s_p are added as backtrack points.

The DPOR algorithm presented in this section performs a stateless exploration. In Chapter 4 it will be further extended to perform a stateful exploration of first acyclic and then arbitrary

---

[8]Thus t_{i-1} is a transition that executes a *delta-* or *timed notification* such that the successor s_i is a state in the *evaluation* phase

---

Algorithm 6: hbrBacktrackAnalysis(s, e)

Input: State and visible effects of the last transition
Output: Backtrack analysis with using a *happens before* relation

1 **procedure** *hbrBacktrackAnalysis(s, e)* **is**
2   **for** $s_p, e_p \in transitionsOfLastDeltaCycle(P)$ **do**
3   **if** *areDependent(e_p, e)* **then**
4   E $\leftarrow$ $\{thread(t) \mid t \in enabled(s_p)\}$
5   **if** *thread(e)* $\in E$ **then**
6   T $\leftarrow$ $\{next(s_p, thread(e))\}$
7   **else if** $\exists e_j$ *after* e_p *with* $e_j \trianglelefteq_P thread(e)$ *and* $thread(e_j) \in E$ **then**
8   T $\leftarrow$ next(s_p, thread(e_j))
9   **else**
10   T $\leftarrow$ enabled(s_p)
11   *addBacktrackPoints(s_p, T)*

---

finite state spaces. The algorithm is separated into multiple functions, so it can be more easily extended. It simplifies the identification of common parts between the different version of the algorithm.

Variations The backtrack analysis function *commonBacktrackAnalysis* of the DPOR algorithm can be replaced with the function *hbrBacktrackAnalysis* presented in algorithm Algorithm 6. The algorithm additionally makes use of the adapted *happens before* relation introduced in Section 2.6.3. It can potentially compute smaller backtrack sets and thus explore less transitions and states. The conducted experiments suggest that maintaining a happens before relation often does not yield significant improvements to the basic algorithm. So it might not be worth the implementation and runtime overhead[9].

The DPOR algorithm does not necessarily perform better (explore less transitions and states) than a SPOR algorithm. However both algorithms can be combined such that a statically computed persistent set is returned when the dynamic algorithm is unable to infer a non trivial persistent set [FG05]. To do so, all occurrences of *enabled(s)*, in the lines $\{5, 7, 30, 26\}$, are replaced with *persistentSet(s)*, where *persistentSet(s)* is an arbitrary persistent set in s. This extension also analogously works with the *hbrBacktrackAnalysis* function. Any one of the algorithms presented in Section 2.6.2 can be used to compute a static persistent set in a state s. Thus backtrack points are constrained to a statically persistent set that can be computed once, when a state is visited the first time.

Example Exploration This section presents a small example program that shall illustrate the above stateless DPOR exploration. The program is available in Listing 2.8. It consists of three threads A,B and C. Each of them has a single transition, thus they will also be simply referred to as A,B and C. Its state space is finite and acyclic, thus all assertion violations will be found by the above DPOR algorithm. The program is unsafe, the transition sequence B,C will violate the assertion in Line 17.

---

[9]The extended backtrack analysis presented here simply assumes that the current happens before relation is available in Line 7. An actual implementation has to compute and maintain such a relation.

```
1  int a = 0;        8               14              20
2  int b = 0;        9  thread B {   15  thread C {   21  main {
3                    10    if (a == 0) {  16    if (b > 0) {  22    start;
4  thread A {        11      b = 1;   17      assert     23  }
5    a = 1;          12    }         18        false;
6    b = 0;          13  }           19    }
7  }                                     }
```

Listing 2.8: Example program to demonstrate the DPOR algorithm

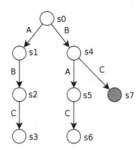

Figure 2.5.: Example scheduling that may be performed by the stateless DPOR algorithm for the program in Listing 2.8

Table 2.1.: Relevant informations when applying the DPOR algorithm to the program in Listing 2.8

step	search path	working	done
0	s_0	$\{s_0: \{A\}\}$	$\{s_0: \emptyset\}$
1	$s_0 \xrightarrow{A} s_1$	$\{s_0: \emptyset,\ s_1: \{B\}\}$	$\{s_0: \{A\},\ s_1: \emptyset\}$
2	$s_0 \xrightarrow{A} s_1 \xrightarrow{B} s_2$	$\{s_0: \emptyset,\ s_1: \emptyset,\ s_2: \{C\}\}$	$\{s_0: \{A\},\ s_1: \{B\},\ s_2: \emptyset\}$
3	$s_0 \xrightarrow{A} s_1 \xrightarrow{B} s_2 \xrightarrow{C} s_3$	$\{s_0: \emptyset,\ s_1: \emptyset,\ s_2: \emptyset,\ s_3: \emptyset\}$	$\{s_0: \{A\},\ s_1: \{B\},\ s_2: \{C\},\ s_3: \emptyset\}$
4	$s_0 \xrightarrow{A} s_1 \xrightarrow{B} s_2$	$\{s_0: \{C\},\ s_1: \emptyset,\ s_2: \emptyset\}$	$\{s_0: \{A\},\ s_1: \{B\},\ s_2: \{C\}\}$
5	$s_0 \xrightarrow{A} s_1$	$\{s_0: \{B,C\},\ s_1: \emptyset\}$	$\{s_0: \{A\},\ s_1: \{B\}\}$
6	s_0	$\{s_0: \{B,C\}\}$	$\{s_0: \{A\}\}$
7	$s_0 \xrightarrow{B} s_4$	$\{s_0: \{C\},\ s_4: \{A\}\}$	$\{s_0: \{A,B\},\ s_4: \emptyset\}$
8	$s_0 \xrightarrow{B} s_4 \xrightarrow{A} s_5$	$\{s_0: \{C\},\ s_4: \emptyset,\ s_5: \{C\}\}$	$\{s_0: \{A,B\},\ s_4: \{A\},\ s_5: \emptyset\}$
9	$s_0 \xrightarrow{B} s_4 \xrightarrow{A} s_5 \xrightarrow{C} s_6$	$\{s_0: \{C\},\ s_4: \emptyset,\ s_5: \emptyset,\ s_6: \emptyset\}$	$\{s_0: \{A,B\},\ s_4: \{A\},\ s_5: \{C\},\ s_6: \emptyset\}$
10	$s_0 \xrightarrow{B} s_4 \xrightarrow{A} s_5$	$\{s_0: \{C\},\ s_4: \{C\},\ s_5: \emptyset\}$	$\{s_0: \{A,B\},\ s_4: \{A\},\ s_5: \{C\}\}$
11	$s_0 \xrightarrow{B} s_4$	$\{s_0: \{C\},\ s_4: \{C\}\}$	$\{s_0: \{A,B\},\ s_4: \{A\}\}$

An example scheduling, that the DPOR algorithm may perform, is shown in Figure 2.5. The states are named in the order they are visited, so s_0 is visited first and s_7 last. Some relevant information that the DPOR algorithm tracks during the execution of the program is available in Table 2.1. It shows the search path together with the *working* and *done* sets for each state on the search path. Already explored transitions are stored in the *done* set and transitions that have been identified during the backtracking are stored in the *working* set for each state. Every row in the table shows a snapshot of these informations. The *step* column is used to identify the snapshots during the following description. Initially all transitions are enabled, thus $en(s_0) = \{A, B, C\}$. Whenever a transition is executed it will be disabled in the successor state.

Initially in step 0, the search path contains only the initial state s_0, all transitions A,B,C are enabled in s_0. The transition A has been selected to be explored first, hence $s_0.working = \{A\}$. Doing so will move the system into step 1. The sets $s_0.working$ and $s_0.done$ have been updated to $\emptyset$ and $\{A\}$ respectively. The last state on the search path is s_1, which is the successor of s_0 when executing transition A. The transition B is selected to be explored from s_1, thus $s_1.working = \{B\}$. Doing so will reach the state s_2 (step 2). The *done* and *working* sets of s_1 are update accordingly. Next the last enabled transition C is selected to be explored from s_2. This will reach the terminal state s_3 (step 3).

Since the last state of the search path s_3 is a terminal state, the backtrack analysis will be triggered. The last transition C is checked for dependencies with the previous transitions. It is independent with B from s_1, since Line 11 has not been executed on this path, but has a *read/write* dependency with A from s_0. The transition C is enabled in s_0 and has not been explored from s_0 so far, thus $s_0.working$ is updated to $\{C\}$. The last state s_3 is removed from the path. Now (step 4) s_2 is the last state and B the last transition. Since $s_2.working = \emptyset$ backtracking continues. So B is compared with all previous transitions. A *read/write* dependency between B and A from s_0 is detected. Hence B is added to $s_0.working$. The last state s_2 is removed from the path. Now (step 5) s_1 is the last transition. Backtracking continues, since $s_1.working$ is also empty. Since no transitions have been executed before A, and A does not disable any transitions itself, no further backtrack points are computed. So s_1 is removed from the path and s_0 becomes the last state (step 6).

Its working set is not empty, so backtracking stops and the normal exploration continues. Now B is selected to be explored resulting in the state s_4 (step 7). Then A is selected reaching s_5 (step 8) and lastly C is executed reaching the terminal state s_6 (step 9). Since the last state is a terminal state, backtracking is applied. The last transition C is compared with the transitions A from s_4 and B from s_0. It turns out that C is dependent with both transitions. Both *working* sets, of s_0 and s_4, are updated to contain C. The working set of s_5 is not updated, since C does not disable any other transition. Now s_6 is removed and s_5 becomes the last state (step 10). Backtracking continues since $s_5.working$ is empty. The last transition A is detected to be dependent with B from s_0. Since $A \in s_0.done$, the *working* set of s_0 is not updated. Next s_5 is removed and s_4 becomes the last state (step 11). Its working set contains only C, thus C is explored from s_4. Doing so will violate the assertion in Line 17 and the search terminates[10].

---

[10]Otherwise it would terminate once the initial state has been backtracked, which means that the complete state space has been explored.

3. Static Partial Order Reduction in Stateful Model Checking

POR is a particularly effective technique to alleviate the state explosion problem by limiting the exploration of redundant transition interleavings. POR works by selecting only a subset of enabled transitions in each step, which are sufficient to prove the desired properties. The other transitions are temporarily ignored, since they are non-interfering with those that are selected. A stateful search on the other hand avoids re-exploration of already visited states. A naive combination of both techniques can lead to the point, where a relevant transition is permanently ignored due to a cycle in the reduced state space. This situation is commonly referred to as (transition/action) *ignoring problem*. It has been first identified by [Val89]. The ignoring problem needs to be solved in order to preserve properties more elaborate than deadlocks. The focus of this thesis is on the verification of safety properties specified in the form of assertions, thus the ignoring problem has to be considered. The solution is to incorporate a so called *cycle proviso* to prevent transition ignoring.

The rest of this chapter is organized as follows: First sufficient conditions will be presented such that a partial order reduced state space exploration preserves all deadlocks and safety properties (assertion violations). Then based on these conditions, concrete implementations of stateful static partial order reduced state space explorations are presented.

3.1. Preserving Properties of Interest

Partial order reduction alleviates the state space explosion problem by limiting the exploration of redundant transition interleavings. For each state s only a subset of enabled transitions $r(s)$ is explored. This results in the exploration of a reduced state space A_R. A reduced state space A_R can be defined in terms of a reduction function r, as described in the preliminaries in Section 2.6.1. The reduction function $r(s)$ cannot be chosen arbitrarily but it has to satisfy certain conditions in order to preserve all properties of interest from A_G. For example choosing $r(s) = \emptyset$ will not preserve any interesting properties of the complete state space A_G, since only the initial state would be reached. This section presents sufficient (non trivial[1]) conditions for the reduction function, in order to preserve deadlocks and safety properties (assertion violations).

3.1.1. Deadlocks

In order to preserve all deadlocks in A_R it is sufficient that the following two conditions hold for every state s in the reduced state space ($s \in S_R$) [God96]:

C_0 $r(s) = \emptyset$ iff $en(s) = \emptyset$

---

[1]A trivial sufficient condition would be that $r(s) = enabled(s)$ for each state $s \in S_R$. Since $A_R = A_G$ in this case, all properties of A_G are clearly preserved. But no reduction has been achieved at all. So the idea is to use a reduction function that will preserve all properties of interest while yielding a *good* state space reduction.

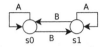

Figure 3.1.: Abstract example state space demonstrating the ignoring problem

C_1 $r(s)$ is a persistent set in s

The first condition ensures that persistent sets are not empty if there are enabled transitions. The second condition ensures that all relevant transition interleavings are explored. The correctness of this method is established by the following theorem.

Theorem 3.1.1

> Let A_R be a reduced state space where the reduction function r satisfies the conditions C_0 and C_1. Let s_d be a deadlock reachable from the initial state s_0 in A_G by a trace w. Then s_d is also reachable from s_0 in A_R.

By Definition 3 of reduced state spaces, the initial state s_0 is always in A_R. The idea is to show that at least one equivalent trace $w' \in [w]_{s_0}$ is explored in A_R. By Definition 6 of equivalent traces, both of them will lead to the deadlock s_d in A_G. The complete proof is available in the Appendix in Section A.2.

3.1.2. Ignoring Problem

The above algorithm will provably reach all deadlock states in A_G but it can miss assertion violations in general. The reason is that some relevant transition of the complete state space A_G may be ignored along some cycle in the reduced state space A_R. Consider the abstract example state space in Figure 3.1. It consists of two states s_0 and s_1 and the transition relation $\Delta = \{(s_0, A, s_0), (s_0, B, s_1), (s_1, B, s_0), (s_1, A, s_1)\}$. Assuming that the transitions A and B are independent, s_0 is the initial state and both transitions are enabled in s_0. Then $r(s_0) = \{A\}$ would be a valid reduction function that satisfies the condition C_0 and C_1. Thus the state s_1 will not be explored, so if it would be an error state, it would be missed.

This problem has already been identified by [Val89] as *(action/transition) ignoring problem*. The idea to solve the problem is to require another condition, called the *cycle proviso/condition*, to ensure that no relevant transition is ignored.

IVL Specific Example Program

A concrete IVL example program that demonstrates the ignoring problem in the context of SystemC is available in Listing 3.1. A complete reduced state space that satisfies the conditions C_0 and C_1 is shown in Figure 3.2. The program consists of three threads A, B and D. Each of them consists of multiple transitions separated by the context switches in line 5, 8, 14 and 20 respectively, so e.g. the thread B consists of two transitions, B_1 (lines 13-14) and B_2 (lines 15-14). The designated transition t_N denotes the execution of the notification phase, which is a SystemC specific simulation detail[2]. A valid dependency relation according to Definition 4 is defined as the symmetric and reflexive closure of:

$$\text{dep} = \{(A_2, B_1), (A_2, B_2), (A_3, B_1), (A_3, B_2)\}$$

---

[2]Explicitly modeling the execution of the notification phase as a transition, prevents the introduction of special cases, when combining SystemC specific details with general state exploration methods.

```
 1  event eA;            10  }                  19  thread D {
 2  event eB;            11                     20    delay 0;
 3                       12  thread B {         21    assert false;
 4  thread A {           13    while (true) {   22  }
 5    delay 0;           14      wait eB;       23
 6    while (true) {     15      notify eA;     24  main {
 7      notify eB;       16    }                25    start;
 8      wait eA;         17  }                  26  }
 9  }                    18
10  }
```

Listing 3.1: IVL example program demonstrating the ignoring problem.

Figure 3.2.: Possible reduced state space for the program in Listing 3.1 satisfying conditions C_0 and C_1.

The following transitions are enabled and selected for each state:

$en(s_0) = \{A_1, B_1, D_1\}$ $\qquad$ $r(s_0) = \{A_1\}$

$en(s_1) = \{B_1, D_1\}$ $\qquad$ $r(s_1) = \{B_1\}$

$en(s_2) = \{D_1\}$ $\qquad$ $r(s_2) = \{D_1\}$

$en(s_3) = \{t_N\}$ $\qquad$ $r(s_3) = \{t_N\}$

$en(s_4) = \{A_2, D_2\}$ $\qquad$ $r(s_4) = \{A_2\}$

$en(s_5) = \{B_2, D_2\}$ $\qquad$ $r(s_5) = \{B_2\}$

$en(s_6) = \{A_3, D_2\}$ $\qquad$ $r(s_6) = \{A_3\}$

$en(s_7) = \{B_2, D_2\}$ $\qquad$ $r(s_7) = \{B_2\}$

Thus clearly the reduced state space defined by the reduction function r satisfies the conditions C_0 and C_1. The transition D_2 is enabled in s_4 but never explored from s_4 or any state reachable from it. Consequently the assertion violation in Line 21 is missed.

Remark. There also exists a reduced state space satisfying C_0 and C_1 and detecting the assertion violation. But in general C_0 and C_1 are not sufficient to enforce this.

3.1.3. Safety Properties

Due to the ignoring problem, a cycle proviso is required to preserve properties more elaborate than deadlocks. Different provisos are available for both safety and liveness properties in combination with different search strategies [God96; HGP92; BH05; BLL06; EP10].

The following proviso from [EP10] yields promising reduction results, is compatible with a DFS and allows to preserve safety properties.

C_2 $\quad$ For every state s in A_R, if $t \in en(s)$ then there is a state s' reachable from s in A_R such that $t \in r(s')$.

Basically it states that any enabled transition is eventually explored in the reduced state space. It is a sufficient condition to prevent the ignoring problem.

Conditions C_0, C_1 and C_2 together are sufficient to preserve safety properties, e.g. specified in form the of assertions, from A_G in A_R. In fact the condition C_0 is no longer necessary (it was

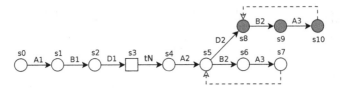

Figure 3.3.: Possible reduced state space for the program in Listing 3.1 satisfying conditions C_1 and C_2.

for deadlocks though). The reason is that C_2 is already strictly stronger and thus implies C_0, as stated by the following Lemma 3.1.2. This can be easily shown by a direct proof, which is available in the Appendix in Section A.6. In the following condition C_0 will still sometimes be explicitly considered even though it is actually unnecessary.

Lemma 3.1.2. *Condition C_2 implies condition C_0.*

Thus the conditions C_1 and C_2 are already sufficient to preserve all assertion violations from A_G in A_R. The correctness of this proposition is established by the following theorem.

Theorem 3.1.3

> Let A_R be a reduced state space where the reduction function r satisfies the conditions C_1 and C_2 as defined in this section. Let w be a trace in A_G that leads to an error state from the initial state s_0. Then there exists a trace w_r in A_R from s_0 such that $w \in Pref([w_r]_{s_0})$.

The idea is to show that for each trace w in A_G from s at least one trace w' is explored in A_R from s such that w is a prefix of w', denoted as $w \in Pref([w']_s)$. The function $Pref$ takes a set of traces W and returns the set of all prefixes defined as $Pref(W) = \{w_p \mid w_p$ is a prefix of any $w \in W\}$. For example let $w' = ACB$ be a trace explored from a state s where the transitions B and C are independent, then the equivalent traces of w' are defined as $[w']_s = \{ACB, ABC\}$. Thus the set of prefixes of equivalent traces of w' is defined as:

$$Pref([w']_s) = \{\emptyset, A, AB, AC, ABC, ACB\}$$

So e.g. $w = ABC \in Pref([w']_s)$ and also $w = AC \in Pref([w']_s)$.

Example 2. Consider again the example program from the previous section demonstrating the ignoring problem in the context of SystemC. The (relevant) transition D_2 is enabled in the states $\{s_4, s_5 = s_7, s_6\}$ but never explored. Condition C_2 ensures that $D_2 \in r(s_j)$ for some $j \in \{5..7\}$. Lets assume that $D_2 \in r(s_5)$. The modified explored state space is shown in Figure 3.3. The trace $w = A_2 B_2 D_2$ leads to an error state in A_G from s_4 (the trace part from s_0 to s_4 is uninteresting for this example). The trace $w' = A_2 D_2 B_2$ is in A_R and $w \in Pref([w']_{s_4})$, since D_2 is independent with B_2. Similarly $D_2 \in Pref([A_2 D_2]_{s_4})$ and $A_2 B_2 A_3 B_2 D_2 \in Pref([A_2 D_2 B_2 A_3 B_2]_{s_4})$. In the same way there exists a trace w' in A_R from state s for every trace w in A_G such that $w \in Pref([w']_s)$.

A stronger condition C_2^S that, in combination with C_1, implies C_2 but might be easier to implement is (adapted from [EP10]):

C_2^S For every state s in A_R there exists a reachable state s' from s in A_R, such that s' is fully expanded, i.e. $r(s') = en(s')$.

This condition states it explicitly that all states $s \in S_R$ that reach a fully expanded state s' are safe from the ignoring problem. For IVL programs it directly implies that cycles of states that span multiple delta cycles do not suffer from the ignoring problem, since a delta cycle can only be left through a state s where only a single designated transition t_N, namely the notification phase transition, is enabled. Thus s is fully expanded. It follows immediately that every state on the cycle with s can reach a fully expanded state. Consequently, the *ignoring problem* is only limited to single delta cycles in the context of SystemC.

The correctness of the C_2^S condition is established by the following theorem. A proof is available in the Appendix in Section A.6.

Theorem 3.1.4

> The condition C_2 is implied by the condition C_1 and C_2^S.

3.2. Static Partial Order Reduced Exploration

This section presents two stateful state space exploration algorithms. Both are based on a DFS and employ SPOR. The first one will preserve all deadlocks. It will be called DPE (Deadlock Preserving Exploration). The second algorithm extends the first one to preserve all assertion violations. It will be called AVPE (Assertion Violations Preserving Exploration). Both algorithms compute static persistent sets during the search. Any one of the algorithms presented in the previous section can be used to this end.

3.2.1. Preserving Deadlocks

The DPE (Deadlock Preserving Exploration) algorithm is shown in Algorithm 7. It is a selective search algorithm, that explores all reachable states s of the reduced state space, defined by the following reduction function $r(s) = selectTransitions(s)$ as:

$$selectTransitions(s) = \begin{cases} \emptyset & \text{if } en(s) = \emptyset \\ persistentSet(s,t) & \text{else, where } t \in en(s) \end{cases}$$

thus it clearly satisfies the conditions C_0 and C_1 for all reachable states s. The function *persistentSet* uses one of the algorithms presented in Section 2.6.2 to compute a persistent set in s starting with an arbitrary enabled transition t in s.

3.2.2. Cycle Proviso

In order to preserve assertion violations in a cyclic state space, the ignoring problem has to be solved. This section describes in general how to integrate the cycle proviso C_2^S into a selective DFS. The idea is to mark states that can reach a fully expanded state (*safe*) or potentially lie on a cycle (*unfinished*). Whenever a state s is about to be backtracked that is unfinished and not safe, which means it can potentially ignore a relevant transition, it will be refined. This means that some additional transitions are explored from s. It will be described with more details in the following.

Every state s is associated with two additional flags $s.safe$ and $s.unfinished$, which initially are set to $false$ for each state. The flags are not considered when comparing two states for equality, so $s_1 = s_2$ can be true, even though $s_1.safe \neq s_2.safe$ or $s_1.unfinished \neq s_2.unfinished$.

Algorithm 7: Stateful Deadlock Preserving Exploration (DPE)

Input: Initial state

1 $H \leftarrow Set()$
2 $explore(\texttt{initialState})$

3 **procedure** $explore(s)$ **is**
4 | **if** $s \notin H$ **then**
5 | | $H.add(\texttt{s})$
6 | | **for** $t \in selectTransitions(s)$ **do**
7 | | | $n \leftarrow succ(s,t)$
8 | | | $explore(\texttt{n})$

The flag $s.safe$ means that a fully expanded state s' is reachable from s. Thus a safe state clearly satisfies the sufficient cycle proviso C_2^S which implies the weaker proviso C_2. Whenever a fully expanded state s' is reached, all states that can reach s' are necessarily $safe$ too. The states currently on the search stack represent a path from the initial state s_0 to s', thus all of them can be marked safe.

The flag $s.unfinished$ means that s potentially lies on a cycle of states. A state s is marked unfinished, if s has already been explored, s is not already marked safe and a state s' is about to be explored such that $s = s'$. If a state s does not lie on a cycle it will eventually reach a terminal state s'. Since s' is fully expanded, s will be marked $safe$. This leads to the point that every state is either safe when backtracked or reaches an unfinished state.

Every state that is marked unfinished and not already marked safe can potentially ignore a relevant transition, so it will be refined when backtracked. The following algorithms simply fully expand a state during refinement. Thus ultimately every explored state can reach a fully expanded state. Consequently the cycle proviso C_2^S is satisfied which implies the weaker proviso C_2.

Incorporating the cycle proviso into the DPE algorithm, results in the exploration of a reduced state space, which is induced by the following reduction function:

$$r(s) = \begin{cases} enabled(s) & \text{if } \neg s.safe \wedge s.unfinished \\ selectTransitions(s) & \text{else} \end{cases}$$

This reduction function satisfies the conditions C_0, C_1 and C_2. But it only is a declarative solution, the flags $s.safe$ and $s.unfinished$, as described above, still need to be computed somehow. The flags for a state s are naturally not available before s is explored. But to explore s, by using the above reduction function $r(s)$, the flags are required. The solution to break up this circular dependency is to separate the exploration of transitions into multiple steps. Since $T = selectTransitions(s) \subseteq enabled(s)$, so the transitions in T would be explored anyway, first all transitions in T are explored and afterwards s is refined if necessary. The following section shows such an implementation.

Remark. As already mentioned, the proviso C_2^S has been introduced in [EP10]. The authors already provide a description on how to integrate the proviso in a DFS. But they assume that it can be decided for states that have not yet been explored whether they are safe or not. Compared to it, this section provides an explicit, instead of a declarative, method to satisfy the cycle proviso in a DFS.

Algorithm 8: Stateful Assertion Violation Preserving Exploration (AVPE)

Input: Initial state

```
 1  H ← Set()
 2  stack ← [initialState]
 3  explore()

 4  procedure explore() is
 5  |   s ← stack.top()
 6  |   if s ∈ H then
 7  |   |   v ← H[s]
 8  |   |   if v.safe then
 9  |   |   |   stack.markAllSafe()
10  |   |   else
11  |   |   |   v.unfinished ← True
12  |   else
13  |   |   H.add(s)
14  |   |   s.safe ← False
15  |   |   s.unfinished ← False
16  |   |   if r(s) = en(s) then
17  |   |   |   stack.markAllSafe()
18  |   |   T ← selectTransitions(s)
19  |   |   exploreTransitions(s, T)
20  |   |   if ¬ s.safe ∧ s.unfinished then
21  |   |   |   stack.markAllSafe()
22  |   |   |   exploreTransitions(s, en(s)\T)

23  procedure exploreTransitions(s, T) is
24  |   for t ∈ T do
25  |   |   n ← succ(s, t)
26  |   |   stack.push(n)
27  |   |   explore()
28  |   |   stack.pop()
```

3.2.3. Preserving Assertion Violations

This section presents a stateful exploration algorithm that employs SPOR and preserves all assertion violations of the complete state space. The algorithm is shown in Algorithm 8. It will be called AVPE (assertion violations preserving exploration). It implements the cycle proviso discussed in the previous section on top of the deadlock preserving algorithm (DPE) presented in Section 3.2.1. Thus it naturally satisfies the conditions C_0, C_1 and C_2. The rest of this section will describe the actual implementation of the AVPE in more details. A further refined algorithm that also explicitly handles multiple successors due to symbolic branches and the different simulation phases of the SystemC runtime is presented in the Appendix in Section A.7.

The AVPE algorithm starts with an empty state cache H. It explicitly manages a search stack, which initially contains the initial state. The algorithm starts by calling the *explore* procedure, which will recursively explore all reachable states.

The *explore* procedure considers the last state s on the current search stack. If s has already been explored, then the already visited state $v \in H$ such that $v = s$ is retrieved. If v is safe, then all states in the search stack must also be safe, since they represent a path from s_0 to $s = v$. The *markAllSafe* procedure marks all of them to be safe. Otherwise v is marked as unfinished. Thus v has eventually to be refined when backtracked. If s has not already been explored, then it will be added to the cache H and the flags safe and unfinished are initialized to be False. Next it will be checked whether the selected subset of transitions, due to POR, equals the set of all enabled transitions. If so, s is fully expanded and can be marked safe. Thus all states that can reach s can also be marked safe. This is done by calling the *markAllSafe* function of the search stack in Line 17. It marks all states currently on the stack as safe. Since s is currently the top of the stack, it will be marked too. Next a subset of enabled transitions T of s is selected in Line 18, as defined for the DPE algorithm in Section 3.2.1. All transitions in T are recursively explored by calling *exploreTransitions* in Line 19. Once the procedure returns, this state s is about to be backtracked. So it is checked in Line 20, whether s can potentially ignore a relevant transition and thus needs refinement. This is the case if s does not necessarily reach a fully expanded state ($\neg s.safe$) and may potentially be on a cycle of states ($s.unfinished$). In this case s will be fully expanded. Therefore s necessarily is safe and all states currently in the stack too, since they lead to s, thus they are all marked safe in Line 21. Consequently, all enabled transitions which have not been explored so far, are recursively explored in this case in Line 22. Finally, and similarly to the previous algorithms presented so far, the cache H will be notified that s is backtracked. A backtracked state is no longer (or immediately will be removed as in this algorithm) on the search stack, thus it cannot form a cycle with any unexplored state currently on the search stack, which makes it a good candidate to free some cache memory, if required, without affecting the correctness of the exploration algorithm as already discussed in Section 2.5.

Remark. The algorithm presented in this section preserves assertion violations in general, independent of the employed algorithm to compute persistent sets. However, under certain conditions the ignoring problem can already be implicitly solved. More information on this topic is available in the Appendix in Section A.9.

4. Dynamic Partial Order Reduction in Stateful Model Checking

This chapter presents a stateful DPOR algorithm that is applicable in the context of SystemC. Preliminaries regarding DPOR are already presented in Section 2.6.3, including basic definitions and a basic stateless algorithm based on [FG05] and [KGG08]. The algorithm is sound, it will preserve all assertion violations, if the state space is finite and acyclic. However, a naive stateful extension is potentially unsound as relevant transition dependencies may be missed during backtracking. This situation will be referred to as the *missing dependency problem*. It can arise in both acyclic and cyclic state spaces. The goal of this section is to extend the basic algorithm to support cyclic state spaces, resulting in the SDPOR (Stateful DPOR) algorithm. To achieve this goal, a step-wise extension is employed:

1. The first section introduces the *missing dependency problem* in acyclic state spaces. Then the stateless algorithm will be extended to a stateful exploration of acyclic finite state spaces, called A-SDPOR (Acyclic SDPOR), in Section 4.2. Thus it is *merely* an optimization that avoids multiple explorations of the same state. The extensions are in principle similar to those proposed in [Yan+08]. However they have to be adapted to the SystemC context.

2. The A-SDPOR algorithm is unsound when applied to cyclic state spaces, since it will no longer compute persistent sets. The problem, which is a variant of the *missing dependency problem*, will be described in Section 4.3. A novel extension of the A-SDPOR that solves the *missing dependency problem* in cyclic state spaces is proposed in Section 4.3. The resulting algorithm is denoted C-SDPOR (Cyclic SDPOR).

3. In general the ignoring problem has to be solved in order to preserve safety properties[1] in cyclic state spaces. The C-SDPOR algorithm, in its standard form, already implicitly solves the ignoring problem in the context of SystemC. Thus an explicit cycle proviso is not required for the standard C-SDPOR algorithm. However all DPOR variants can fall back to static persistent sets, whenever the inference of non trivial persistent sets fails dynamically. Doing so will make the C-SDPOR susceptible to the ignoring problem. For this reason the C-SDPOR algorithm is combined with the cycle proviso C_2, as already defined in Section 3.1.3 and further discussed in Section 3.2.2, in Section 4.4 to solve the ignoring problem in general, independent of the method used to compute persistent sets[2]. The resulting algorithm is denoted simply as SDPOR, since it is the final version.

---

[1] And also liveness properties, or in general all properties more elaborate than deadlocks.

[2] Combining the DPOR algorithm with static persistent sets in general requires the integration of a cycle proviso. The DPOR algorithm itself could use a more sophisticated dynamic analysis to infer persistent sets, which could yield a smaller number of trivial persistent sets (fully expanded state) and thus require an explicit cycle proviso.

```
1  int a = 0;          8                    14                    20
2  int b = 0;          9  thread B {        15  thread C {        21  main {
3                      10    if (a == 0) {  16    if (b > 0) {     22    start;
4  thread A {          11      b = 1;       17      assert        23  }
5    a = 1;            12    }              18    false;
6    b = 0;            13  }               18    }
7  }                   13 }                19  }
```

Listing 4.1: Example program that appeared in the preliminaries as Listing 2.8 to demonstrate the basic DPOR algorithm

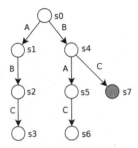

Figure 4.1.: Example scheduling that may be performed by the stateless DPOR algorithm for the program in Listing 4.1

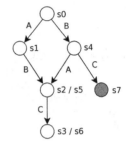

Figure 4.2.: Relevant part of the complete state space for the exploration of the program Listing 4.1 following the execution order as shown in Figure 4.1

4.1. Missing Dependency Problem

The program in Listing 4.1 is used to illustrate the missing dependency problem. It has already been used in the preliminaries Section 2.6.3 as an example to illustrate the stateless DPOR state space exploration. The program is displayed here again for convenience. The explored state space is shown in Figure 4.1. It is only a part of the complete state space, since the transition sequence B,C leads to an error state from the initial state s_0, thus the exploration aborts early. The complete state space is finite and acyclic. The relevant part of the complete state space is shown in Figure 4.2. It explicitly shows that the explored states s_2 and s_5 are completely equal. So actually it is unnecessary to explore both of them (and with it all of their successors). Thus it is desirable to combine DPOR with a stateful exploration, even if the state space has no cycles. But it turns out that a stateful extension of DPOR is non trivial even for finite acyclic state spaces [FG05; YWY06; Yan+08].

For example consider the following straightforward implementation. A global set H is managed to store already visited states. Whenever a new state s is reached it will first be checked whether s has already been visited. If not s is added to H and the search normally continues. Else the search path is aborted and normal backtracking starts from s. These changes could easily be integrated into the basic stateless DPOR algorithm shown in Algorithm 5 by placing the check $s \in H$ in Line 4 and only continue in the *explore* function if the check returns false. But such an implementation is potentially unsound, as it can miss assertion violations. This effect can also be observed in the same example program in Listing 4.1.

Assuming that the transitions will be executed in the same order, the extended version will work exactly the same as the stateless version until the state s_5 is reached (since $s \in H$ would return false up to this point). The current search path is $s_0 \xrightarrow{B} s_4 \xrightarrow{A} s_5$. It is detected that $s_5 = s_2 \in V$ and backtracking starts with s_5 as the last state. Consequently, the transition C, which would have been executed from s_5 and has been executed from s_2, is not considered by the backtrack analysis. Thus the dependency of C with A from s_4 is missed and C is not added to the $s_4.working$ set, which in turn does not lead to the exploration of C from s_4 and the assertion violation in Line 17 of the program is missed.

4.2. Stateful Exploration of Finite Acyclic State Spaces

This section provides a stateful extension, denoted as A-SDPOR (Acyclic Stateful DPOR) and shown in Algorithm 9, to the basic DPOR algorithm that preserves all deadlocks and assertion violations in a finite acyclic state space. The extensions provided here are similar to those provided in [Yan+08], adapted to the SystemC context. The idea is to keep a record of the visible effects of every executed transition. Whenever the currently explored state s is matched with an already visited state v (Line 7), an extended backtrack analysis will be initiated and the exploration of s stops. The extended analysis will collect all visible effects of transitions executed from v or any states reachable from v and use them to check for dependencies with the visible effects associated with the transitions on the current path P. The next section describes the algorithm in more details. Then an example is provided and finally the correctness of this extension is briefly discussed.

A-SDPOR Algorithm

The A-SDPOR algorithm is an extension to the basic stateless Algorithm 5. The modifications are shown in Algorithm 9. Functions, which are not shown there, are inherited from the basic algorithm without changes. Changes in existing functions are underlined for convenience. The A-SDPOR algorithm keeps a cache H to avoid re-exploration of already visited states and a directed graph G_V to keep a record of the visible effects for every executed transition. Whenever a transition t is executed from state s leading to the successor state s', an edge $(id(s), id(s'))$ is added to G_V, where $id(s)$ is some unique identifier for s. The edge is associated with the visible effects e, that have been tracked during the execution of t from s Line 23.

The states itself are not stored in G_V but only some unique identifier (e.g. a unique number or the memory address of the state, depending on the actual implementation). Storing the complete state would result in a significant memory overhead. Only the visible effects are stored which are collected during the execution of the corresponding transition. These are normally relatively small, since they only contain simple identifiers, e.g. memory locations that have been accessed or events that have been immediately notified.

In the following the transition itself will also be displayed alongside the source and target state for convenience. So edges will be written as (s, t, s') instead of (s, s') even though the transition t can be unambiguously obtained, given the source and target states (s, s'). The notation $E(s, t, s')$ returns the visible effects, which were observed when executing transition t from s reaching state s'. The corresponding transition and its thread can be obtained from the visible effects e by $transition(e)$ and $thread(e)$ respectively.

Whenever the currently explored state s is matched with an already visited state v (Line 7), an extended backtrack analysis will be initiated and the exploration of s stops (so afterwards a normal backtrack analysis will be applied). The extended analysis will lookup v in G_V and find all visible effects of transition that can be executed from v or any of its successor states.

Algorithm 9: A-SDPOR: Stateful DPOR for Acyclic State Spaces

Input: Initial state

```
1  H ← Set()
2  Gᵥ ← VisibleEffectsGraph()
3  P ← Trace(initialState)
4  explore()

5  procedure explore() is
6      s ← last(P)
7      if s ∈ H then
8          v ← H[s]
9          extendedBacktrackAnalysis(v)
10     else
11         H.add(s)
12         if enabled(s) ≠ ∅ then
13             s.done ← {}
14             s.working ←
                   {selectAny(enabled(s))}
15             while |s.working| > 0 do
16                 exploreTransitions(s,
                       s.working)
```

```
17 procedure exploreTransitions(T) is
18     for t ∈ T do
19         s.working ← s.working \ t
20         s.done ← s.done ∪ t
21         n ← succ(s, t)
22         e ← visibleEffects(s, t)
23         Gᵥ.addEdge(id(s), id(n), e)
24         P.push(e, n)
25         explore()
26         P.pop()
27         normalBacktrackAnalysis(s, e)

28 procedure extendedBacktrackAnalysis(v) is
29     for e ∈
           Gᵥ.reachableTransitionEffects(id(v)) do
30         commonBacktrackAnalysis(e)
```

For example if $G_V = [(s_0, A, s_1), (s_1, B, s_2), (s_2, C, s_3)]$ and $v = s_1$ the call in Line 29 would collect the reachable effects $\{E(s_1, B, s_2), E(s_2, C, s_3)\}$. The visible transition effects can than be used to compute dependencies with transitions (and their collected visible effects during their execution) of the current search path.

Example

For example consider again the example program in Listing 4.1. Due to the missing dependency problem, which has been introduced in the last section, a naive stateful extension could miss the assertion violation in Line 17. Lets assume that this acyclic SDPOR algorithm has reached state s_5, by first executing $s_0 \xrightarrow{A} s_1 \xrightarrow{B} s_2 \xrightarrow{C} s_3$, then backtracking to s_0 and executing $s_0 \xrightarrow{B} s_4 \xrightarrow{A} s_5$, which is now the current search path P. Up to this point it behaved exactly as the stateless DPOR algorithm, since $s_5 = s_2$ is the first state that has already been visited. Thus the current state $s = s_5$ and the already visited state $v = s_2$. The visible effects graph G_V currently contains the following edges: $[(s_0, A, s_1), (s_1, B, s_2), (s_2, C, s_3), (s_0, B, s_4), (s_4, A, s_5)]$ So for this example the extended backtrack analysis would start at $v = s_2$ and collect all reachable edges, which are in the same delta cycle, since the dependency analysis does not proceed across delta cycles in the DPOR algorithm. In this case only a single edge will be returned, namely (s_2, C, s_3). The effects of all collected edges are used to compute dependencies with transitions currently in the search path P. Thus the dependency of C from s_2 with A from s_4 is detected and C is added to the $s_4.working$ set, which in turn will eventually lead to the exploration of C from s_4 and the assertion violation in Line 17 of the program is detected correctly.

Remark. The extended backtrack starts with an already visited state v and collects a set E of all reachable visible effects. All effects $e \in E$ are then checked for dependencies with transitions (actually their visible effects) on the current search path P. The order in which the $e \in E$ are considered does not matter. The result will be the same.

Notes on Correctness

Assuming that the DPOR algorithm explores persistent sets for each state s, i.e. $s.done$ is a persistent set for each state s when the search is completed, it follows immediately that the A-SDPOR algorithm explores persistent sets too when applied to a finite acyclic state space. If no state s is matched with an already visited state v, then DPOR and A-SDPOR behave completely identically. Otherwise, a state s is matched with an already visited state v. The state v has already been backtracked and thus is no longer on the search path P, else the state space would be cyclic which it is not by assumption. Consequently, all relevant transitions from v have already been explored and their visible effects recorded. Since v is completely equal with s the effects generated from v are exactly the same than those that would be generated from s. And since the state space is finite and acyclic by assumption, they are exactly the same a stateless DPOR would yield. Consequently, if the basic stateless DPOR algorithm is correct (in the sense that it preserves all deadlocks and assertion violations), than this stateful extension will also be. The stateless algorithm may visit the same transition multiple times, since the same state may be reached multiple times on different paths. The stateful extension on the other hand will explore each state and each of its transitions at most once.

4.3. Supporting Finite Cyclic State Spaces

The previous sections so far have considered the combination of DPOR with stateful exploration for finite acyclic state spaces. Now the focus will shift to arbitrary finite state spaces. The reason for the step-wise extension is that it is more easy to support acyclic state spaces and the same extensions can be reused as foundation for a support of arbitrary finite state spaces.

In the following, first the problem with the A-SDPOR algorithm, when applied to cyclic state spaces, is outlined by an example. Then an extension of the A-SDPOR is proposed, called C-SDPOR (Cyclic SDPOR), to solve the problems. Finally some notes on the correctness of this extension are given. A complete proof is not yet available though, thus the soundness of this extension has not been formally established and is left for future work. Thus C-SDPOR might *only* perform a under-approximation, so it can be used for bug finding, in cyclic state spaces. Nevertheless, C-SDPOR has yielded the same result as the provably correct AVPE algorithm, which is based on static persistent sets, on all conducted experiments.

4.3.1. Missing Dependency Problem in Cyclic State Spaces

The A-SDPOR algorithm collects all, for dependency analysis, relevant informations for every executed transition, called the visible effects of each transition. Whenever a state s is matched with an already visited state v, all visible effects of every transition reachable from v are analyzed with the current path leading to s, since $s = v$ would explore exactly the same transitions which would have exactly the same visible effects. It works in acyclic state spaces, since v must already have been fully explored, thus all relevant visible effects reachable from v are available. This is not the case in cyclic state spaces. Since v itself would be on the path, the informations for v are not yet complete. Thus the A-SDPOR algorithm could miss relevant dependencies for transitions on the path between the last state s and the already visited state v.

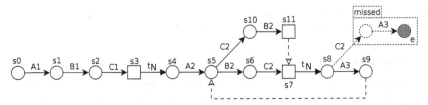

Figure 4.3.: Example state space that the A-SDPOR algorithm could explore for the program in Listing 4.2. The designated transition t_N denotes the execution of a notification phase as defined in Section 2.4.1. Thus they separate the delta cycles of the simulation. The dotted states and transitions are relevant to detect the error state e, but they are missed by the A-SDPOR algorithm in this exploration (and thus in general). Matched states are connected by dashed lines with unfilled arrows, where the target is an already explored state currently in the cache.

An IVL program that demonstrates this missing dependency problem is shown in Listing 4.2. It consists of three threads and they in turn consist of multiple transitions, denoted as A_1, A_2 and so on, separated by context switch statements (all wait O statements). The program is unsafe, as the transition sequence $A_1 B_1 C_1 t_N A_2 B_2 C_2 t_N C_2$ will lead to an error state. The designated transition t_N denotes the execution of a notification phase, which is a SystemC specific simulation detail, which transfers the simulation from one delta cycle to the next.

A possible state space exploration, which the A-SDPOR would perform, is available in Figure 4.3. The states are visited in the order they are numbered. So the path $s_0 \xrightarrow{A_1} s_1 \xrightarrow{B_1} s_2 \xrightarrow{C_1} s_3 \xrightarrow{t_N} s_4 \xrightarrow{A_2} s_5 \xrightarrow{B_2} s_6 \xrightarrow{C_2} s_7 \xrightarrow{t_N} s_8 \xrightarrow{A_3} s_9$ is explored first. It will be detected that $s_9 = s_4$ has already been visited. Thus the A-SDPOR algorithm will initiate the extended backtrack analysis. Starting from $v = s_4$ all (distinct) reachable effects will be collected, that are in the same delta cycle as $v = s_4$, these are: $[(s_5, B_2, s_6), (s_6, C_2, s_7)]$, and analyzed with the transitions executed in the last delta cycle of the current path P for dependencies, these are: $[(s_8, A_3, s_9)]$. Neither the transition B_2 from s_5 nor C_2 from s_6 writes the global c variable, thus there exist no dependencies with A_3 from s_8 which asserts that the value of c is zero (thus reads c). Since no dependencies are detected, no backtrack points are added. Even though there exists a feasible path namely $s_8 \xrightarrow{A_3} s_9 = s_5$ and then $s_5 \xrightarrow{C_2} s_{10} \xrightarrow{B_2} s_{11}$ such that C_2 from s_5 is read/write dependent with A_3 from s_8. So actually C_2 should be added to $s_8.working$. But it does not happen when the A-SDPOR algorithm is used. The problem is that C_2 has not yet been explored from s_5, so it is not considered during this dependency analysis. It will only be explored once backtracking reaches the state s_5 [3]. Consequently, C_2 will not be explored from s_8 and the assertion violation in Line 8 is missed.

Remark. Executing C_2 from s_5 in this example will result in a write access to the global variable c, thus a read/write dependency with A_3 from s_8 would be available. The execution of C_2 from s_6 on the other hand resulted only in a write of the global variable b, thus it was not dependent with A_3.

---

[3]The dependency between B_2 from s_5 and C_2 from s_6 will be detected later on during normal backtracking. Since C_2 is enabled in s_5 it will then be added to $s_5.working$. Once backtracking does reach the state s_5, C_2 will be explored from s_5. But at this point s_8 has already been backtracked and is no longer available on the search path.

```
1   int b=0;                13   while (true) {       26        if (b > 0) {
2   int c=0;                14     wait_time 0;       27           b = 0;
3                           15     if (c > 0) {       28        } else {
4   thread A {              16       c = 0;           29           c = 1;
5     wait_time 0;          17     } else {           30        }
6     while (true) {        18       b = 1;           31     }
7       wait_time 0;        19     }                  32   }
8       assert (c == 0);    20   }                    33
                            21 }                      34   main {
9   }                       22                        35     start;
10  }                       23   thread C {           36   }
11                          24     while (true) {
12  thread B {              25       wait_time 0;
```

Listing 4.2: Example program to demonstrate that the A-SDPOR algorithm is unsound for cyclic state spaces

---

Algorithm 10: C-SDPOR Extensions and Overrides of the A-SDPOR Algorithm 9

---

1 **procedure** *explore()* **is**
2     $s \leftarrow$ last(P)
3   **if** $s \in H$ **then**
4        $v \leftarrow H[s]$
5   **if** $v \in P$ **then**
6           $c \leftarrow$ collectStateCycle(v)
7   expandStateCycle(c)
8   **else**
9   *extendedBacktrackAnalysis(v)*

10   **else**
11   *H*.add(s)
12   s.done $\leftarrow \{\}$
13   **if** $en(s) \neq \emptyset$ **then**
14   s.working $\leftarrow \{$selectAny(enabled(s))$\}$

15   **while** $|s.working| > 0$ **do**
16   *exploreTransitions*(s, s.working)

17 **procedure** *collectStateCycle(v)* **is**
18     $c \leftarrow \{\}$
19     $S \leftarrow$ P.statesLastDeltaCycleReversedOrder()
20   **for** $s \in S$ **do**
21        $c \leftarrow c \cup s$
22   **if** $s = v$ **then**
23   **break**

24   **return** c

25 **procedure** *expandStateCycle(c)* **is**
26   **for** $s \in c$ **do**
27   *addBacktrackPoints*(s, *enabled*(s))

4.3.2. C-SDPOR Algorithm

The main reason for the *missing dependency problem* in cyclic state spaces is that, when a cycle of states is detected between the currently explored state s and an already visited state v, not all information relevant for backtracking are available yet for v, since v itself is on the current path P. Thus the A-SDPOR algorithm could miss relevant dependencies for transitions on the path between the last state s and the already visited state v.

This problem has also been identified in [YWY06]. The proposed solution is to compute a summary of interleaving information (SII) for each state, basically which transitions can be explored from each state, and update it as necessary, since the SII is incomplete for states on a cycle. Apart from the rather complex implementation, this adds additional runtime overhead and memory requirements to store and update the SII for each state. Though it has been noted in [YWY06] that the additional memory requirements are rather insignificant compared to the requirements of storing states in a stateful exploration and traditional methods such as state compression can be used to reduce the memory requirements.

In this thesis instead a simple novel solution is proposed to solve the mentioned problem for cyclic state spaces. Whenever a cycle is detected, all states of the cycle, which have been executed in the last delta cycle of the simulation, are fully expanded. Doing so cannot miss any relevant transition, since all transitions[4] are explored. At first glance this seems to be a solution that does not scale, but if the number of state cycles and the number of relevant states (those in the last delta cycle) per cycle is not too large, the method should perform sufficiently good. Also many of the redundant interleavings that are introduced when a cycle of states is detected can be pruned by a stateful search. The reason is that many of them are non-interfering and thus will lead to the same state. For example consider the (partial) abstract state space in Figure 4.4, which contains a cycle $s_1 \overset{A}{\to} s_2 \overset{B}{\to} s_3 \overset{C}{\to} s_1$. If e.g. the transition D is non-interfering with A in s_1, then the transition sequences $s_1 \overset{A}{\to} s_2 \overset{D}{\to} s_5$ and $s_1 \overset{D}{\to} s_4 \overset{A}{\to} s_5$ will lead to the same state s_5, thus only one of them will be further pursued. Furthermore, as already mentioned, backtracking does not progress across delta cycles in the context of SystemC, thus only those states on the cycle are fully expanded, that belong to the last delta cycle. This can reduces the number of fully expanded states significantly. For the example state space shown in Figure 4.3 only a single state, namely s_8, out of the four states ($s_5 = s_9, s_6, s_7, s_8$) on the cycle, will be expanded. Finally, the algorithm can also fall back to static persistent sets, as the stateless DPOR and A-SDPOR algorithm, instead of fully expanding a state. Whether or not a more sophisticated method, which could impose considerable implementation complexity and runtime overhead, is worth the additional effort is to be investigated. Preliminary experiments (Chapter 7) show sufficiently good results for the simple algorithm.

The extended algorithm will be called C-SDPOR, it extends the algorithm provided in Algorithm 9, thus all available functions are inherited, and overrides the explore function as shown in Algorithm 10. Whenever the currently explored state s is matched with an already explored v, it will be checked whether v is currently in the search path P. If the check in Line 5 yields true, all states in P, starting from v until s form a cycle of states (including v, excluding s). This cycle will be collected in Line 6 and than expanded in Line 7. Only states in the last delta cycle are included into the collected cycle of states, since dependencies are not propagated across delta cycles. Else the path is acyclic. Thus an extended backtrack analysis will be applied in Line 9 as the A-SDPOR algorithm would do it.

As already mentioned, the algorithm can also fall back to static persistent sets, as the stateless DPOR and A-SDPOR algorithm. In addition to the changes necessary for the DPOR algorithm, as described in Section 33, the *expandStateCycle* function would restrict the backtrack points in Line 7 to a statically computed persistent set, instead of adding all enabled transitions.

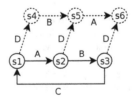

Figure 4.4.: Part of an abstract state space

---

[4]All of the last delta cycle, since the DPOR algorithm does not backtrack across delta cycles in the context of SystemC, as already mentioned in Section 2.6.3.

4.3.3. Notes on Correctness

A partial order reduced state space exploration of cyclic state spaces has to handle the ignoring problem, as introduced in Section 3.1.2, in order to preserve more elaborate properties than deadlocks. The C-SDPOR algorithm naturally solves the ignoring problem. Since the state space is finite, every path w from every state s will either reach a terminal state or run into a cycle. A terminal state by definition is fully expanded. By definition of the C-SDPOR algorithm, for every cycle of states, at least one state is fully expanded. Thus from every state of the cycle a fully expanded state is reachable. So in either case s will reach a fully expanded state. Thus the cycle proviso C_2^S, which implies the weaker condition C_2, is satisfied by the C-SDPOR algorithm.

It is left to show that the C-SDPOR algorithm satisfies condition C_1, which states that $r(s) = s.done$ is a persistent set when s is finally backtracked. In Section 30 it has been reasoned that the A-SDPOR algorithm explores persistent sets for each state in a finite acyclic state space. The idea has been to reduce it to the basic stateless DPOR algorithm. The C-SDPOR works identically to the A-SDPOR algorithm for acyclic state spaces. So it is left to show that the C-SDPOR algorithm explores persistent sets for each state in a finite cyclic state space. A formal proof (or disprove by counterexample) of this property is beyond the scope of this thesis and left for future work.

When C-SDPOR is configured to fall back to static persistent sets, denoted as combined C-SDPOR, than it no longer does satisfy the condition C_2, so it suffers from the ignoring problem. In fact the same example program shown in Listing 3.1 in Section 3.1.2 that demonstrates this problem for SPOR will also produce the problem for the combined C-SDPOR. Thus it makes sense to also further extend the C-SDPOR algorithm to explicitly handle the ignoring problem independently of implementation details on how persistent sets are inferred.

4.4. Complete Stateful Exploration

This section extends the C-SDPOR algorithm such that the ignoring problem is solved in general independent of the backtracking analysis or whether static persistent sets are used as fall back whenever the inference of non trivial persistent sets fails dynamically. This extended algorithm will simply be called SDPOR (stateful DPOR), as it is the final extension.

The SDPOR algorithm is completely equal to the C-SDPOR algorithm except for the *explore* procedure. The adapted *explore* function is available in Algorithm 11. The basic idea to implement the DFS compatible cycle proviso C_2^S has already been described in Section 3.2.2, when extending the deadlock preserving exploration (DPE) to an assertion violations preserving exploration (AVPE). The same principles also apply here. The idea is to mark states that can reach a fully expanded state (safe) or potentially lie on a cycle (unfinished). Whenever a state s is about to be backtracked that is unfinished and not safe, which means it can potentially ignore a relevant transition, s will be fully expanded. Thus ultimately every explored state can reach a fully expanded state. Consequently the cycle proviso C_2^S is satisfied which implies the weaker proviso C_2. In the following the extended *explore* procedure, which implements the cycle proviso, will be described in more details. Finally some notes about the correctness of the SDPOR algorithm will be given. A complete implementation of SDPOR, that falls back to static persistent sets, is available in the Appendix in Section A.8.

Algorithm 11: SDPOR Extensions of the C-SDPOR Algorithm 10

---

1 **procedure** *explore()* **is**
2 | s ← last(P)
3 | **if** $s \in H$ **then**
4 | | $v \leftarrow H[s]$
5 | | **if** *v.safe* **then**
6 | | | *P.markAllSafe()*
7 | | **else**
8 | | | v.unfinished ← True
9 | | **if** $v \in P$ **then**
10 | | | c ← collectStateCycle(v)
11 | | | expandStateCycle(c)
12 | | **else**
13 | | | *extendedBacktrackAnalysis(v)*
14 | **else**
15 | | H.add(s)
16 | | s.done ← {}
17 | | **if** *enabled(s)* $\neq \emptyset$ **then**
18 | | | s.working ← {*selectAny(enabled(s))*}
19 | | | **while** $|s.working| > 0$ **do**
20 | | | | *exploreTransitions*(s, s.working)
21 | | **if** *s.done = enabled(s)* **then**
22 | | | *P.markAllSafe()*
23 | | **else if** $\neg$ *s.safe* $\wedge$ *s.unfinished* **then**
24 | | | *P.markAllSafe()*
25 | | | *exploreTransitions*(s, *enabled(s)* \ *s.done*)

The Extended Explore Procedure

The explore function considers the last state s on the current search path P. If s has already been explored, then the already visited state $v \in H$ such that $v = s$ is retrieved. If v is safe, then all states in P must also be safe, since they represent a path from s_0 to $s = v$. Else v is marked as unfinished. Thus v has eventually to be refined when backtracked.

If s has not already been explored, then the SDPOR algorithm first works identical to the C-SDPOR algorithm. The state s is added to the state cache. Then it will be initialized. During initialization every newly reached state will be associated with a unset safe and unfinished flag. Then identically to the C-SDPOR algorithm, an arbitrary enabled transition will be selected and explored. This can lead to the detection of backtrack points due to dependencies between the executed transitions. These are stored in the *working* set of s. The loop will iterate until no further backtrack points are computed. The C-SDPOR algorithm would be finished at this point. The SDPOR algorithm checks whether the current state s has to be refined to circumvent the ignoring problem.

If all enabled transitions have been explored (Line 21), it means that s is fully expanded, all states in the current search path P (including s) are marked as *safe*. The reason is that P represents a path from the initial state s_0 to s, thus clearly every state on the path can reach a fully expanded state, namely s. Else it will be checked whether s has to be refined (Line 23). This is the case if s is part of a cycle (unfinished) and does not reach a fully expanded state (not safe). Refining a state means that additional transitions are explored from that state. The current implementation simply fully expands s, by exploring all enabled transitions that have not already been explored. Consequently s and all states leading to it can be marked safe (Line 24).

Notes on Correctness

As discussed above, this SDPOR algorithm satisfies the cycle proviso condition C_2. This leaves to show that condition C_1 is satisfied too. The SDPOR algorithm performs completely equal to the C-SDPOR algorithm until an *unsafe* and *unfinished* state s is backtracked. In that case s will be fully expanded. Thus $s.done$ will be a trivial persistent set. For all other states, persistent sets are computed with the same reasoning as for the C-SDPOR algorithm, as described in the previous section. If C-SDPOR satisfies condition C_1, so does SDPOR.

5. State Subsumption Reduction

A stateful search keeps a set of already visited states to avoid re-exploration of the same state. Whenever a new state s is reached, it will be matched with the already visited states. The basic stateful algorithm presented in Section 2.5 requires two states to be fully equal in order to be matched. However, in the context of symbolic execution, where states consists of concrete and symbolic state parts, this is an unnecessary strong condition. The reason is that symbolic expressions can be represented in many (actually infinitely) different ways, e.g. the expressions $a + a$ and $2 * a$ are structurally different (hence not equal) but semantically equivalent. Furthermore, a path condition (pc) is managed during symbolic execution. It represents constraints that the symbolic expressions have to satisfy, thus effectively limiting the possible values a symbolic expression can assume. In general a symbolic expression represents a set of concrete values. A symbolic state thus represents sets of concrete states. If the set of concrete states represented by a symbolic state s_2 contains the set of concrete states represented by a symbolic state s_1, it is not necessary to explore s_1 if s_2 has already been explored. For example consider two states s_1 and s_2, that are equal except for a variable v and their path conditions, which are defined as: $v_1 = 2 * x$, $pc_1 = x \geq 3 \wedge x \leq 7$ and $v_2 = x + 5$, $pc_2 = x \geq 0 \wedge x \leq 10$ respectively. The symbolic variable v_1 represents the set of concrete values $\{6, 8, 10, 12, 14\}$, which is a subset of the values $\{5..15\}$ represented by v_2, in combination with their corresponding path conditions. It is unnecessary to explore s_1, if s_2 has already been explored, since s_2 subsumes all possible behaviors of s_1. State s_1 is said to be *subsumed/covered* by s_2 in this case, denoted as $s_1 \preccurlyeq s_2$. The example shows, that detecting subsumption between states can further increase the state matching rate, thus reducing the exploration of redundant states, compared to equivalence detection, which would miss the reduction opportunity outlined in the example. This reduction technique will be called State Subsumption Reduction (SSR).

Additionally, subsumption is a general concept that is not limited to symbolic values. It can also be considered as a generalization of state equivalence and also applied to concrete state parts to further increase the number of state matches. For example, it will be shown that under certain conditions states can be matched with different simulation times. Furthermore complementary techniques like heap- and process- (thread) symmetry reduction, e.g. as described in [Ios02], can also be considered a form of state subsumption. These techniques are not further considered in this thesis, since threads are (currently) not dynamically created, thus the threads and their heap references are already normalized.

The rest of this chapter begins by introducing necessary definitions. This includes a specification of the state (subsumption) matching predicate $\preccurlyeq$ and a formalization of execution states. Then in Section 5.2 and Section 5.3 sufficient conditions will be presented, such that SSR can be combined with basic model checking and POR, while preserving assertion violations. Finally a concrete implementation (algorithm) of the predicate $\preccurlyeq$ is provided in Section 5.4. The algorithm will first compare the concrete state parts, and if they match, then the symbolic parts will be compared. An exact method is used to check subsumption of symbolic values.

5.1. Definitions

State Subsumption Reduction (SSR) can, similarly to Partial Order Reduction (POR), be defined in terms of a reduced state space $A_R = (S_R, s_0, T, \Delta_R, s_e)$. A reduced state space A_R can be obtained from another state space $A = (S, s_0, T, \Delta, s_e)$ (e.g. the complete state space A_G) by a reduction function r, as described in Section 2.6.1.

Definition 13 (*State Subsumption Reduction Function*)

$$r(s) = \begin{cases} \emptyset & \text{if } \exists s' \in S_R : s_0 \xrightarrow{*}_R s' \land s \preccurlyeq s' \land r(s') = en(s') \\ en(s) & \text{else} \end{cases}$$

Basically exploration of a state s is avoided, if a state s' has already been explored with $s \preccurlyeq s'$. The states s and s' will be called *similar* or *matching* in such a case. The state s is said to be *subsumed* or *covered* by s'. The effectiveness of the reduction greatly depends on the accuracy of the $\preccurlyeq$ predicate. Defining $s \preccurlyeq s'$ as $s = s'$, i.e. complete equality, will yield no reduction at all, thus $A_R = A_G$ in this case. On the other hand it can obviously not be chosen arbitrarily in order to preserve all properties of interest. For example simply returning *True* for all state pairs will result in a reduced state space where only the transitions of the initial state are explored. All the successors s_i of the initial state s_0 would be detected to be matched with s_0, hence $r(s_i)$ would be the empty. So far $\preccurlyeq$ has been used intuitively. The following formal definition of $\preccurlyeq$ is suitable to preserve assertion violations.

Definition 14 (*Result Coverage/Subsumption*)

A state s_1 is (result) covered/subsumed by a state s_2, denoted as $s_1 \preccurlyeq s_2$, iff for every trace w that leads to an error state from s_1, w also leads to an error state from s_2.

The binary relation $\preccurlyeq$ defines a partial order on the state space. If $s_1 \preccurlyeq s_2$ and s_2 has already been explored, then it is not necessary to explore s_1, in order to detect all safety property violations. Result coverage can be extended to result equivalence.

Definition 15 (*Result Equivalence*)

A state s_1 is (result) equivalent to a state s_2, denoted as $s_1 \simeq s_2$, iff $s_1 \preccurlyeq s_2$ and $s_2 \preccurlyeq s_1$.

The binary relation $\simeq$ defines an equivalence relation on the state space. By definition it immediately follows that equivalence reduced state space contains all states and transitions of the coverage reduced state space. Equivalence is strictly stronger but a predicate deciding equivalence might be easier to implement than one deciding coverage/subsumption. If either $s_1 \simeq s_2$ or $s_1 \preccurlyeq s_2$ then the states s_1 and s_2 are said to be *similar* or *matching*.

It is useful to adapt the notion of reachability for such reduced state spaces, since often states s are reached that have enabled transitions but are not explored, because a similar state has already been visited. The following definition of reachability, denoted as *weak reachability*, defines the execution of traces that allows to express paths more naturally when state subsumption matching by means of the $\preccurlyeq$ predicate is used.

Definition 16 (*Weak Reachability*)

Let s and s' be states in A_R. The state s' is weakly reachable from the state s by a (possibly empty) trace $w = t_1..t_n$ in A_R iff $s \preccurlyeq s'$ or there exists a state s'' in A_R such that

- $s \preccurlyeq s''$ and

- $t_1 \in r(s'')$ and

- let s_t be the successor of s'' when executing t_1, then s' is weakly reachable from s_t by the rest trace $t_2..t_n$ of w.

The states s, s' and s'' do not have to be different, since $s \preccurlyeq s$ holds for all states s. Thus every state s is weakly reachable from itself by the empty trace.

Let s and s' be states in A_R, s' is weakly reachable from s in A_R iff there exists a trace w such that s' is weakly reachable from s by w, as defined above.

Similar to the normal reachability, weak reachability is also a reflexive and transitive relation over the state space. Weak reachability is a generalization of normal reachability. Thus for all two states s_1, s_2 it holds that, if s_2 is reachable from s_1, then s_2 is also weakly reachable from s_1. The following example shall illustrate the notion of weak reachability. An abstract example state space is shown in Figure 5.1. The subsumption matching relation is defined as:

$$\preccurlyeq = \{(s_9, s_6), (s_7, s_3)\}$$

Thus $s_9 \preccurlyeq s_6$ and $s_7 \preccurlyeq s_3$ holds. The reachability relation R is the smallest transitive reflexive closure satisfying:

$$R = \{(s_0, s_1), (s_1, s_2), (s_2, s_3), (s_3, s_4), (s_0, s_5), (s_5, s_6), (s_6, s_7), (s_0, s_8), (s_8, s_9)\}$$

Where (a, b) denotes that b is reachable from a. The weak reachability relation R_W is the smallest transitive reflexive closure satisfying $R \subseteq R_W$ and:

$$\{(s_9, s_6), (s_7, s_3)\} \subseteq R_W$$

Thus e.g. s_4 is weakly reachable from s_8 by trace CAD, which goes through the states $s_8, s_9, s_6, s_7, s_3, s_4$. And s_7 is weakly reachable from s_0 by trace CBA or BCA, which go through the states s_0, s_5, s_6, s_7 or s_0, s_8, s_9, s_6, s_7 respectively.

5.1.1. Execution State

Each execution path is represented with an execution state and the other way round each execution state corresponds to an execution path. Thus both terms may be used interchangeably. An execution state (or just state) can formally be defined as:

Definition 17 (*Execution state*)

$State := (time, pc, vars, notifications, threads = [ts_1, ..., ts_n])$
$ts_i := (name, loc, await, status)$

Where *time* is the current simulation, *pc* denotes the path condition and *vars* is a name to value mapping of all variables. For simplicity it is assumed, that the names of global and local variables are disjunct. Thus a single mapping is sufficient to manage all variables. The mapping *notifications* maps time deltas to events, so once a time delta reaches zero (in the delta or timed

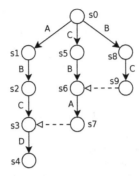

Figure 5.1.: Abstract state space illustrating the concept of weak reachability

notification phase), its associated event is triggered. The list *threads* contains the state (ts ≡ thread state) of each thread. It consists of the thread name (any unique identifier will do), a location *loc*, one can think of it as program counter, a status and a wait condition *await*, which is either the special value *None* or an event, that this thread awaits to become enabled. The status of a thread is either: Enabled, Disabled or Terminated. A thread becomes terminated when it finishes the execution of all its statements. A thread is disabled when it waits for an event or is suspended. Else it is enabled. The variable values and path conditions are allowed to be symbolic. All other state parts are concrete.

Remark. Timed notifications are not tracked in the state, since they can be reduced to normal events. Each statement wait_time n; can be equivalently rewritten as a sequence of statements { notify e, n; wait e; } ,where *e* is a unique event, as described in Section 2.1.

A mapping *m* (vars and notifications are modeled as mappings) consists of a sequence of items. Each item is a (key,value) pair. Every key can only appear once: $\forall (k_1, v_1), (k_2, v_2) \in m : k_1 = k_2 \implies v_1 = v_2$. Each mapping *m* provides the operations *keys*=$\{k \mid (k, \_) \in m\}$ and *values*=$\{v \mid (\_, v) \in m\}$. Additionally the value of a mapping *m* can be retrieved using the corresponding key *k* as $get(m, k) = v$. This operation is defined iff $(k, v) \in m$. It will often be abbreviated as $m[k]$. A mapping, e.g. $m = \{(a, 1), (b, 2)\}$, will often be simply written as $\{a = 1, b = 2\}$ or just $\{a: 1, b: 2\}$ to improve readability.

The operation *linkedValues*(m_1, m_2)=$\{(v_1, v_2) \mid (k_1, v_1) \in m_1 \land (k_2, v_2) \in m_2 \land k_1 = k_2\}$ is defined on two mappings m_1 and m_2 and returns a sequence of corresponding value pairs, value which are stored under the same key, of both mappings.

Remark. The actual implementation uses states that are a bit more complex then the one described here. So for example it manages a callstack for each thread. Each callframe contains a local program counter and a mapping of local variables. Also variable values are treated as atomar entities in the state description provided here. Actually (variable) values can be more complex compound data structures like classes, arrays and pointers. But all of them can be recursively compared and matched (following the pointer references will lead to another value, classes and concrete arrays just group multiple values, symbolic arrays can be specifically modeled as atomar entities and stackframes contain named slots of variables). Thus the same principles apply for them. But for the following discussion it is irrelevant, whether function calls and complex objects are supported or not. Omitting those details simplifies the following presentation. The implementation however supports these constructs.

Algorithm 12: Check two expressions for equality

Input: Two expressions e_1 and e_2
Output: True iff e_1 equals e_2, i.e. $e_1 = e_2$

1 **function** *matchExpressions(e_1, e_2)* **is**
2   **if** isPrimitive*(e_1)* $\wedge$ isPrimitive*(e_2)* **then**
3      $\lfloor$ **return** $e_1 = e_2$
4   **if** getOperator*(e_1)* $\neq$ getOperator*(e_2)* **then**
5      $\lfloor$ **return** *False*
6   **return** $\forall a1, a2 \in$ zip*(*getOperands *(e_1)*, getOperands *(e_2))* : matchExpressions*(a_1, a_2)*

The set V_S denotes all symbolic literals and the set V_C all concrete values. The combination $P = V_S \cup V_C$ denotes the set of primitive values. A primitive value is either a bitvector or a boolean value. The set Op denotes all (common) arithmetic and logic operators available in the C++ language, e.g. $\{+, *, \&\&, !\} \subseteq Op$. The (infinite) set of expressions E can be recursively defined. Every primitive value is an expression and applying an operator to a set of expressions returns a new expression. Basically E is the set of expressions that can be constructed from symbolic and concrete primitive values, combined with the available operators. Every expression $e \in E$ has a type. An expression is called *primitive* if it only consists of a primitive value. Else it consists of an operator and a non-empty argument list. Such an expression will be called *compound*. The predicates *isPrimitive(e)* and *isCompound(e)* can be used to test an expression. The functions *getOperator(e)* and *getOperands(e)* return the operator or the (ordered) tuple of operands respectively, for the compound expression e. Expressions with the same operand always have the same number of arguments. The function *zip* can be used to pair two tuples of arguments, since arguments are ordered. It is defined as $zip(args_1, args_2) = \{(a_1, a_2) \mid a_1 = args_1[i]$ and $a_2 = args_2[i]$ for $i \in \{1..|args_1|\}\}$, where $args[i]$ returns the i-th element of the arguments tuple. Both primitive and compound expressions can naturally be recursively compared by equality, e.g. $x_1 + 2 = x_1 + 2$ and $x_1 + 2 \neq 2 + x_1$ and $x_1 \neq x_2$ and so on, where $x_1, x_2 \in V_S$ are symbolic literals. Thus $e_1 = e_2$ iff *matchExpressions(e_1, e_2)* returns *True* as defined in Algorithm 12.

Definition 18 (*Structural Compatibility ($s_1 \sim s_2$)*)

Two states s_1 and s_2 are structurally compatible, denoted as $s_1 \sim s_2$, iff threads(s_1) = threads(s_2) and notifications(s_1) = notifications(s_2) and time(s_1) = time(s_2) and vars(s_1) $\sim$ vars(s_2) and pc(s_1) $\sim$ pc(s_2).

vars(s_1) $\sim$ vars(s_2) iff keys(vars(s_1)) = keys(vars(s_2)) and $\forall (v_1, v_2) \in$ linkedValues(vars(s_1), vars(s_2)) : $v_1 \sim v_2$.

Two (primitive) values a,b are structurally compatible ($a \sim b$) iff they may be equal. If a and b are both concrete they may be equal iff $a = b$. If one of them is symbolic, they may be equal if they have the same type.

So basically $s_1 \sim s_2$ iff all concrete state parts are completely equal and symbolic state parts may be equal. A necessary condition that they may be equal is to have the same structure. Thus all variable values of structurally compatible states can be paired. And they will continue with the same statements when resumed (though they can deviate from then on, since variable values and path conditions are not restricted and thus execution of the states can take different branches).

Remark. If $s_1 \sim s_2$ and neither s_1 nor s_2 contains any symbolic state parts then $s_1 = s_2$.

The unary predicates *isSymbolic* and *isConcrete* denote whether a value is symbolic or concrete respectively. The function $symbolicVars(s) = \{v \mid v \in values(vars(s_1)) \text{ and } isSymbolic(v)\}$ returns all symbolic variable values in the state s. It can be naturally extended to work on two states, as the following definition shows:

Definition 19 (*Symbolic Pairs*)

$$symbolicPairs(s_1, s_2) = \{(v_1, v_2) \mid linkedValues(v_1, v_2)$$
$$\wedge (isSymbolic(v_1) \vee isSymbolic(v_2))\}$$

Basically $symbolicPairs(s_1, s_2)$ returns a sequence of corresponding variable values, where at least one variable is symbolic. Normally $s_1 \sim s_2$ is checked beforehand to ensure that all variables can be paired. The notation $var_i(s_1)$ and $var_i(s_2)$ refers to v_1 and v_2 respectively of the i-th pair $(v_1, v_2) \in symbolicPairs(s_1, s_2)$.

The notation $\Gamma(e)$ returns all symbolic literals contained in the expression e, e.g. if $e = x_1 + 2*x_2$ then $\Gamma(e) = \{x_1, x_2\}$. It is also naturally defined on states, thus $\Gamma(s)$ returns all symbolic literals available in any symbolic expression of the state s, i.e. $\bigcup_{v \in symbolicVars(s)} \Gamma(s) \cup \Gamma(pc(s))$.

In the following the focus will be the comparison of symbolic state parts of s_1 and s_2. The assumption is, that s_1 and s_2 are already structurally compatible (else it would not be necessary to compare the symbolic state parts, as the states can't be equal anyway). A symbolic state part of s_1 with respect to s_2 is defined as: $\zeta_{s_2}(s_1) = (pc, vars)$, where $pc = pc(s_1)$ and $vars = \{(k, v) \mid (k, v) \in vars(s_1) \text{ and } (isSymbolic(v) \text{ or } isSymbolic(vars(s_2)[k]))\}$. $\zeta_{s_1}(s_2)$ is defined analogously. Thus the symbolic state part of s_1 depends on the state of s_2. All variables where either one corresponding value is symbolic will be included (the variable name k of s_1 can be used to access variables from s_2, since $s_1 \sim s_2$ by assumption here). For this reason the symbolic part of a state is defined with respect to another state.

Both $\zeta_{s_2}(s_1)$ and $\zeta_{s_1}(s_2)$ have the same number of entries. Their variable names are equal, since $s_1 \sim s_2$ by assumption, but their values (normally) are not. Sometimes $\zeta_{s_2}(s_1) = (pc, vars)$ will be written as $\zeta_{s_2}(s_1) = (pc=pc(s_1), k_1=v_1, ..., k_n=v_n)$ or as $\zeta_{s_2}(s_1) = (pc: pc(s_1), k_1: v_1, ..., k_n: v_n)$ to improve the readability. The second form, that uses a colon as delimiter, might be easier to recognize correctly when the path condition contains equality constraints itself. The prefix $pc =$ or $pc :$ might be completely omitted, since the path condition will always appear first. Here k_i are variable names and v_i are their values, so $(k_i, v_i) \in vars(\zeta_{s_2}(s_1))$. Normally it is clear from the context that s_1 shall be compared with s_2, thus $\zeta_{s_2}(s_1)$ will be simply written as $\zeta(s_1)$ and analogously $\zeta(s_2)$. The following example shall illustrate the previous definitions.

Example State Definitions

This section shows some complete states and their symbolic parts, which would arise when simulating the program in Listing 5.1. It is not mandatory for the understanding of the following concepts but can serve as a concrete reference if some definitions remain unclear.

For convenience, the state template is repeated here again: $State:=(time, pc, vars, notifications, threads = [ts_1, ..., ts_n])$ with $ts_i:=(name, loc, await, status)$. The initial state s_0 is formally defined as:

$s_0 = (0, T, \{a: x_1, b: 0, e: e_1\}, \{\}, threads_0)$
where $threads_0 = [(A, 1, None, Enabled), (B, 1, None, Enabled), (C, 1, None, Enabled)]$

```
1   int a = ?(int);          10                            19
2   int b = 0;               11   thread B {              20   thread C {
3   event e;                 12     if (b != 0) {         21     notify e, 1;
4                            13       b = ?(int) + 2;     22   }
5   thread A {               14     }                     23
6     a += 1;                15     assume (b > 0);       24   main {
7     b += 1;                16     wait e;               25     start;
8     wait e;                17     assert (b > 0);       26   }
9   }                        18   }
```

Listing 5.1: Example program to illustrate the state definitions

The states s_1 and s_2 that are reached from the initial state by executing the transition sequences A,B,C and B,A,C respectively as $s_0 \xrightarrow{A,B,C} s_1$ and $s_0 \xrightarrow{B,A,C} s_2$. They are formally defined as:

$$s_1 = (0, x_2 + 2 > 0, \{a: x_1 + 1, b: x_2 + 2, e: e_1\}, \{1: e_1\}, threads_1)$$
where $threads_1 = [(A, 3, e_1, Disabled), (B, 5, e_1, Disabled), (C, 2, None, Terminated)]$

$$s_2 = (0, T, \{a: x_1 + 1, b: 1, e: e_1\}, \{1: e_1\}, threads_2)$$
where $threads_2 = [(A, 3, e_1, Disabled), (B, 5, e_1, Disabled), (C, 2, None, Terminated)]$

The path condition of s_2 has not been extended, since the condition $\neg(b > 0)$ is unsatisfiable because b has a concrete value in this case. The values of the thread locations (program counters) are somewhat arbitrary. Here just the line number offset to the beginning of the thread is used. Any value that is sufficient to continue the threads execution correctly would be fine here.

The states s_1 and s_2 are structurally compatible, i.e. $s_1 \sim s_2$, since their concrete parts are equal. Their symbolic state parts with respect to each other are defined as: $\zeta(s_1) = (x_2 + 2 > 0, \{a: x_1 + 1, b: x_2 + 2\})$ and $\zeta(s_2) = (T, \{a: x_1 + 1, b: 1\})$. Which can also be equivalently written as: $\zeta(s_1) = (pc: x_2 + 2 > 0, a: x_1 + 1, b: x_2 + 2)$ and $\zeta(s_2) = (pc: T, a: x_1 + 1, b: 1)$. Neither s_1 nor s_2 is structurally compatible with s_0 since their concrete state parts are not equal, e.g. the pending notification of event e is missing.

The function $symbolicPairs(s_1, s_2)$ returns $\{(a_1, a_2), (b_1, b_2)\}$ where a_1, a_2 and b_1, b_2 refer to the corresponding variable values of a and b in s_1 and s_2 respectively. Then $var_1(s_1) = a_1 = x_1 + 1$ and $var_1(s_2) = a_2 = x_1 + 1$ and $var_2(s_1) = b_1 = x_2 + 2$ and $var_2(s_2) = b_2 = 1$.

5.2. Stateful Model Checking with State Subsumption Reduction

A basic stateful model checking algorithm explores the complete state space A_G. Combining it with SSR, results in the exploration of a reduced state space A_R, defined by the reduction function r as shown in Definition 13. For convenience the definition is repeated here:

$$r(s) = \begin{cases} \emptyset & \text{if } \exists s' \in S_R : s_0 \xrightarrow{*}_R s' \land s \preccurlyeq s' \land r(s') = en(s') \\ en(s) & \text{else} \end{cases}$$

Please notice that the function en in the above definition refers to the complete state space A_G and the $\preccurlyeq$ predicate has been defined in Definition 14. The correctness of this combination is established by the following theorem.

Theorem 5.2.1

> Let w be a trace in the complete state space A_G that leads to an error state s_e from the initial state s_0. Let A_R be a corresponding reduced state space using the above reduction function. Then an error state s'_e will also be weakly reachable in A_R from s_0 by w such that $s_e \preccurlyeq s'_e$.

The proof of this theorem is available in the Appendix in Section A.4. Since the subsumption/coverage predicate $\preccurlyeq$ yields state spaces which are a subset of those yield by the equivalence predicate $\simeq$, which has been defined in Definition 15, the correctness of the equivalence predicate is also covered.

5.3. Combination with Partial Order Reduction

In Section 3.1.3 two sufficient conditions C_1 and C_2 have been presented that a state space reduction function r should satisfy in order to preserve all assertion violations. They are suitable to combine POR with a stateful search where states are matched by complete equality. These conditions are sufficient, but in general they are unnecessarily strong. The combination of SSR and POR does not satisfy them. This section presents adapted conditions that support state matching by subsumption, thus preserving assertion violations while combining POR and SSR. In fact, only C_2 has to be adapted, C_1 is still required as is. The idea is to employ the notion of weak reachability, as defined in Definition 16.

C_1 $r(s)$ is a persistent set in s.

C_{2W} Let s be a state in A_R. Let $w = t_1..t_n$ be a non-empty trace leading to an error state from s in A_G. Then there exists a *weakly reachable* state s' from s in A_R, such that at least one transition t_i of w is in $r(s')$.

Please notice that the first transition t_1 of w is enabled in s, i.e. $t_1 \in en(s)$, since by assumption w leads to an error state in A_G, which implies w is executable in A_G. Requiring the reachability of error states for the condition C_{2W} is due to the definition of the state subsumption relation $\preccurlyeq$, shown in Definition 14, since $s_1 \preccurlyeq s_2$ does not require that $en(s_1) \subseteq en(s_2)$. Consequently, a stateful state space exploration algorithm using the $\preccurlyeq$ relation to match states could e.g. not guarantee that $r(s') \neq \emptyset$ if $s \preccurlyeq s'$ and s cannot reach an error state in A_G. So basically condition C_{2W} prevents the ignoring problem for transitions that can lead to an error state. This is a reasonable formulation, since transitions that (provably) cannot lead to an error state can safely be ignored.

Condition C_{2W} is strictly weaker than condition C_2, which were formulated in Section 3.1.3 during the presentation of partial order reduction. So C_2 implies C_{2W} and there exists complete state spaces, where the former conditions will yield a smaller reduction (explore a larger state space) than the latter conditions.

A reduced state space A_R that satisfies C_1 and C_{2W} does preserve all assertion violations of the corresponding complete state space A_G. This result immediately follows from the following theorem.

Theorem 5.3.1 (*Assertion Violations Preserving Combined Reduction*)

> Let A_R be a reduced state space where the reduction function r satisfies the conditions C_1 and C_{2W} as defined in this section. Let w be a trace in A_G leading to an error state from the initial state s_0. Then there exists a trace w_r in A_R such that an error state is weakly reachable from s_0 in A_R.

A proof of this theorem is available in the Appendix in Section A.5. The conditions C_1 and C_{2W} are sufficient to preserve assertion violations in the reduced state space. However, it might be difficult to implement the condition C_{2W}. Thus a stronger condition C_{2W}^S is proposed. In combination with C_1, it implies C_{2W} but might be easier to implement.

C_{2W}^S For every state s in A_R there exists a *weakly reachable* state s' from s in A_R, such that s' is fully expanded, i.e. $r(s') = en(s')$.

The correctness of the C_{2W}^S condition is established by the following theorem:

Theorem 5.3.2

The condition C_{2W} is implied by the conditions C_1 and C_{2W}^S.

A proof of this theorem is available in the Appendix in Section A.6.

5.3.1. Integration of SSR into the AVPE Algorithm

The previous sections have provided sufficient conditions for a state space exploration algorithm that combines SSR with stateful model checking and POR in order to preserve all assertion violations[1].

This section discusses the necessary adaptions to integrate SSR into the AVPE Algorithm 8 from Section 3.2.3, which combines SPOR with stateful model checking. It turns out that only minimal adaptions are necessary. In fact, only the implementation of the check $s \in H$ in Line 6 needs to be modified. It should return *True* iff there exists a state v in H, such that $s \preccurlyeq v$. Thus subsumption matching is used instead of complete equality for every pair of states.

In the following the correctness of this modification is briefly discussed, that the modified AVPE algorithm satisfies the weaker condition C_{2W} and also C_1, thus preserves all assertion violations of the complete state space A_G.

Condition C_1 is not violated by the modification. For each state $r(s)$ is either fully expanded, empty or a persistent set by definition of the AVPE, because transitions are only explored from Line 19 or Line 22. Thus a trivial or non trivial persistent set is explored from every state s in either case.

The cycle proviso condition C_{2W} is satisfied similarly to the cycle proviso C_2 for the normal AVPE algorithm. A state s will be marked safe if a fully expanded state s' is weakly reachable from s. It will be marked unfinished, if it lies on a cycle of weakly reachable states. Whenever a state s is backtracked it will be checked whether it has to be refined. This is the case if s is marked unfinished but not marked safe. In this case s will simply be fully expanded. Thus all states that lie or can reach the cycle of weakly reachable states with s can (weakly) reach a fully expanded state.

5.3.2. Integration of SSR into the SDPOR Algorithm

In order to combine SPOR with SSR, which by definition is a stateful state exploration approach, only the state matching check ($s \in H$) has to be adapted to match states by using the $\preccurlyeq$ predicate instead of complete equality. The combination of DPOR with SSR requires some further adaptions and assumptions about the state matching predicate $\preccurlyeq$. In particular two problems have been identified that can lead to an under-approximation when SSR is directly incorporated into the SDPOR algorithm, which has already been presented in Section 4.4 in Algorithm 11. In the following both problems will be presented and a solution is proposed. The resulting algorithm seems to perform a sound reduction, though a correctness proof is beyond the scope of this thesis. In the *worst* case it performs an under-approximation, since both POR and SSR only explore states of the complete state space A_G, so it can be used for bug finding.

---

[1]Sufficient conditions to preserve deadlocks can be defined similarly.

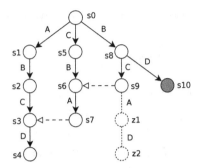

Figure 5.2.: Abstract state space to demonstrate problems that can lead to unsoundness when combining DPOR with SSR. The states z_1 and z_2 are, among others, not explored due to SSR. The error state s_{10} can be potentially missed.

Nonetheless, On all conducted experiments, the SDPOR+SSR combination provided the same result as the SPOR+SSR combination.

First Problem

Consider the abstract example state space in Figure 5.2. The states are visited in the order they are numbered. The designated states z_1 and z_2 will, among others, not be explored due to SSR. The SDPOR algorithm could miss the error state s_{10} by not exploring D from s_8.

Assume that s_9 is the currently explored state. So the paths $s_0 \xrightarrow{A} s_1 \xrightarrow{B} s_2 \xrightarrow{C} s_3 \xrightarrow{D} s_4$ and $s_0 \xrightarrow{C} s_5 \xrightarrow{B} s_6 \xrightarrow{A} s_7$ have already been explored and $s_0 \xrightarrow{B} s_8 \xrightarrow{C} s_9$ is the current search path P. It has already been detected that $s_7 \preccurlyeq s_3$ and s_9 is now matched with s_6, so $s_9 \preccurlyeq s_6$ holds (the check $s = s_9 \in H$ in Line 3 returns *True*). Since s_9 and s_6 do not form a cycle, the SDPOR algorithm will apply an extended backtrack analysis starting from the already visited state $v = s_6$. The visible effects graph G_V contains the following edges: $[(s_0,A,s_1), (s_1,B,s_2), (s_2,C,s_3), (s_3,D,s_4), (s_0,C,s_5), (s_5,B,s_6), (s_6,A,s_7), (s_0,B,s_8), (s_8,C,s_9)]$. So only the effect (s_6,A,s_7) will be checked with the current path P for dependencies. The effect (s_3,D,s_4) is missed entirely since there is no path in G_V from s_6 to s_3. The reason is that s_7 is not necessarily equal to s_3, but only the weaker condition $s_7 \preccurlyeq s_3$ holds. Thus the dependency of C and D in s_8 is missed and the erroneous trace $s_0 \xrightarrow{B} s_8 \xrightarrow{D} s_{10}$ is not explored.

A solution for this problem is to adapt the edges in the visible effects graph whenever a state s is matched with an already visited state v. Since s is explored the first time, there is only a single edge leading to s in G_V, which has been recorded right before the recursive explore call for s. So in Line 4 (of the SDPOR Algorithm 11), right after retrieving the already visited state v, the edge target should be replaced from s to v, e.g. by inserting the method call $G_V.replaceEdgeTarget(from = id(s), to = id(v))$, which performs this replacement. As already noted, only identifiers of states are stored and not the states themselves, so actually the single edge whose target is $id(s)$ is retrieved and its target is replaced by $id(v)$. For this example the edges (s_6,A,s_7) and (s_8,C,s_9) would be replaced with (s_6,A,s_3) and (s_8,C,s_6). So starting from s_6 the effect (s_3,D,s_4) would also be collected and checked with C from s_8 for dependencies. Consequently D would be added to $s_8.working$ and the error state s_{10} would be detected.

Second Problem

Normally, without SSR, the SDPOR algorithm would (eventually) explore the complete path $s_0..z_2$ and then start backtracking. So C from s_8 would be compared with D from z_1. A dependency would be detected in this case and D would be added to the *working* set of s_8. With SSR, C from s_8 is compared with D from s_3, as already described above. The dependency still has to be detected, else the error state s_{10} is missed.

Since the state subsumption relation is preserved by executing the same transition sequence, as shown in Section A.4 in the Appendix, $z_1 \preccurlyeq s_3$ holds. The assumption $z_1 \preccurlyeq s_3$ does not necessarily imply that the dependency of D with C in s_8 is detected. Thus given two states s_1 and s_2 the requirement $s_1 \preccurlyeq s_2$ is not sufficient for the SDPOR algorithm to match s_1 with s_2. Additionally the following requirements are proposed:

D_1 Every trace that can be executed from s_1 in A_G can also be executed from s_2 in A_G.

D_2 $\forall t \in enabled(s_1) : VE(s_1,t) \subseteq VE(s_2,t)$.

Where $VE(s,t)$ returns the visible effects observed during the execution of transition t from state s. Let $e_1 = VE(s_1,t)$ and $e_2 = VE(s_2,t)$ be visible effects. The effect set e_1 is included in e_2, written as $e_1 \subseteq e_2$, if $e_1.isDepenedent(e_3)$ implies $e_2.isDepenedent(e_3)$ for all visible effects e_3. Normally D_1 should already follow from $s_1 \preccurlyeq s_2$, but it is not necessary, so it is explicitly listed. It has already been shown that $s_1 \preccurlyeq s_2$ is preserved for the successor states, whenever the same transition is executed from s_1 and s_2. Analogously the requirement D_1 is preserved too. It can be shown that the algorithms presented in Section 5.4.2 and Chapter 6 to decide state subsumption $s_1 \preccurlyeq s_2$ for states s_1, s_2 already satisfy the additional requirements D_1 and D_2 for the combination of SDPOR and SSR.

5.4. State Matching

A stateful search needs to decide whether a state has already been visited to avoid re-exploration. This process is called *state matching*. As specified in the definitions in Section 5.1.1, states can be divided into a concrete and a symbolic part. Comparing the concrete parts of two states s_1 and s_2 is conceptually and computationally simple. But comparing the symbolic parts is more complex, since semantic equivalent expressions can be syntactically (structurally) different. If the concrete state parts already mismatch, then the states do not match. Thus it makes sense to match the concrete state parts first and only then attempt to match the symbolic state parts. The next section describes how concrete state parts are matched, thus it is used to find candidate states which can match. Then an exact method is presented to detect subsumption between symbolic state parts. Finally Section 5.4.3 shows that states can be equivalent, even though their simulation times are different. This is an example where equivalence between concrete state parts, since the simulation time is (in the current implementation) a concrete value, is exploited to further reduce the explored state space.

5.4.1. Matching Concrete State Parts

A generic algorithm to check whether a state s has already been visited is shown in Algorithm 13. Such an algorithm normally is employed in any stateful model checking method. Visited states are stored in a (hash) set H. To check whether a state s is equivalent to any visited state $s_v \in H$, one would first compute the hash value of s, then lookup all states in H

Algorithm 13: Check whether a state $s_v \in H$ similar to s has already been visited.

Input: State s and hash map H of visited states
Output: True if a *similar* state to s has already been visited, else False

1 **foreach** $s_v \in H[hash(s)]$ **do**
2 | **if** $s \sim s_v$ **then**
3 | | **if** $\zeta(s) \simeq \zeta(s_v)$ **then**
4 | | | **return True**

5 **return False**

---

that have the same hash value, denoted as $H[hash(s)]$. Normally, only the concrete state parts are hashed, because it is not easy to devise a hash function that would preserve the property $hash(s) \neq hash(s_v) \implies s \neq s_v$ with sufficient precision (without filtering out too many states s_v in advance that would be matched with s when compared properly) for symbolic state parts, since they do not have a canonical representation. Then s is compared with all $s_v \in H[hash(s)]$ one after another. If $s \sim s_v$, so their concrete parts are equal, then their symbolic parts are compared with one of the algorithms that will be presented in the following sections. If their symbolic state parts are (detected to be) equivalent, then the states s and s_v are equivalent. To detect subsumed/covered states it would be sufficient to replace the condition $\zeta(s) \simeq \zeta(s_v)$ in Line 3 with $\zeta(s) \preccurlyeq \zeta(s_v)$ (so basically swap the symbolic state part matching algorithm). The Definition 14 and Definition 15 are naturally transferred to symbolic state parts.

Using a hash map is just an implementation detail, albeit a common one. It would also be possible to store all visited states in e.g. a list though it would be less efficient. Common optimization techniques like using multiple hash function for better filtering or incremental hashing to speed up the computation of the hash function can also be used instead of using a single hash value that is fully recomputed each time it is required[2].

5.4.2. Exact Symbolic Subsumption

This section presents an exact method, called Exact Symbolic Subsumption (ESS), to decide subsumption of symbolic state parts. ESS assumes that the compared states s_1 and s_2 are structurally compatible, i.e. their concrete state parts are equal, denoted as $s_1 \sim s_2$, as defined in Section 5.1.1. The idea is to show that every value that can be produced in state s_1 can also be produced in state s_2. A quantified formula that naturally encodes this requirement is:

$$\left(\exists x_1..x_n : pc(s_1) \wedge \bigwedge_{i \in \{1..k\}} var_i(s_1) = v_i \right) \implies \left(\exists y_1..y_m : pc(s_2) \wedge \bigwedge_{i \in \{1..k\}} var_i(s_2) = v_i \right)$$

As already mentioned both states are structurally compatible by assumption ($s_1 \sim s_2$), thus they have the same number of (primitive) variables all of which have the same type. Therefore the variables can be matched, i.e. $var_i(s_1)$ and $var_i(s_2)$ refer to corresponding variables from both states for $i \in \{1..k\}$. The terms $v_1..v_k$ are fresh symbolic literals corresponding to the types of the variables of s_1 and s_2 respectively, i.e. $type(v_i) = type(var_i(s_i)) = type(var_i(s_i))$ for $i \in \{1..k\}$. The symbolic literals $x_1..x_n$ and $y_1..y_m$ are all symbolic literals that appear in either

---

[2]At least for the hashing of concrete state parts, since as already mentioned symbolic state parts are not included into the computation of the hash value due to the lack of a canonical representation.

variable or path condition of state s_1 or state s_2 respectively, i.e. $\Gamma(s_1) = \{x_1,..,x_n\}$ = and $\Gamma(s_2) = \{y_1,..,y_m\}$. With the following auxiliary definition:

$$f(s_1,s_2) = symbolicTriples(s_1,s_2) = \{(v_1,v_2,v_f) \mid (v_1,v_2) \in \texttt{symbolicPairs}(s_1,s_2) \text{ and}$$
$$type(v_1) = type(v_f) \text{ and } v_f \text{ is a fresh symbolic literal}\}$$

The exact symbolic subsumption can be equivalently defined as (making the correspondence between $var_i(s_1)$ and $var_i(s_2)$ explicit):

Definition 20 (*Exact Symbolic Subsumption (ESS)*)

A state s_1 is covered/subsumed by a state s_2, i.e. $s_1 \preccurlyeq s_2$ according to Definition 14, if $ESS_\preccurlyeq(s_1,s_2)$ is valid. $ESS_\preccurlyeq(s_1,s_2)$ is defined as:

$$\left(\exists x_1..x_n : pc(s_1) \wedge \bigwedge_{\substack{(v_1,\_,v_f) \\ \in f(s_1,s_2)}} v_1 = v_f \right) \implies \left(\exists y_1..y_m : pc(s_2) \wedge \bigwedge_{\substack{(\_,v_2,v_f) \\ \in f(s_1,s_2)}} v_2 = v_f \right)$$

where $\{x_1..x_n\} = \Gamma(s_1)$ and $\{y_1..y_m\} = \Gamma(s_2)$ are all symbolic literals that appear in either variables or path condition in state s_1 or s_2 respectively.

In order to show that $ESS_\preccurlyeq$ is valid, it can be shown that $\neg ESS_\preccurlyeq$ is unsatisfiable. The formula $\neg ESS_\preccurlyeq(s_1,s_2)$ can be equivalently rewritten as:

$$\neg ESS_\preccurlyeq(s_1,s_2) = \left(\exists x_1..x_n : pc(s_1) \wedge \bigwedge_{\substack{(v_1,\_,v_f) \\ \in f(s_1,s_2)}} v_1 = v_f \right) \wedge \neg \left(\exists y_1..y_m : pc(s_2) \wedge \bigwedge_{\substack{(\_,v_2,v_f) \\ \in f(s_1,s_2)}} v_2 = v_f \right)$$

Which will be satisfiable if there exists an assignment of variable values that can be produced in state s_1 but not in s_2. If $\neg ESS_\preccurlyeq$ is unsatisfiable then $ESS_\preccurlyeq$ is valid which means s_1 is subsumed by state s_2. Any SMT solver with support for quantifiers can be used to check these quantified formulas for satisfiability. In this thesis the Z3 solver is used.

A correctness proof, that $s_1 \preccurlyeq s_2$ holds if this method detects that s_1 is subsumed by s_2, is available in the Appendix in Section A.10. The idea is to show that every program line that can be reached from state s_1 can also be reached from state s_2, thus exploring s_2 alone is sufficient to detect all assertion violations[3].

Example

For example consider the symbolic state parts $s_1=$(pc: $x_1 \neq 0$, a: $2 * x_1$, b: x_2) and $s_2=$(pc: T, a: $y_1 + 1$, b: $y_2 + y_3$). All symbolic literals reachable from s_1 are $\{x_1, x_2\}$ and from s_2 are $\{y_1, y_2, y_3\}$. Both states (actually their symbolic parts) have two variables a and b, denoted as a_1,b_1 and a_2,b_2 respectively. Both of them have bitvector type, e.g. 32 bit width[4]. Thus two fresh symbolic literals v_1 and v_2 will be introduced with corresponding types for a and b, i.e.

---

[3]An assertion `assert c;` can always be rewritten as conditional statement `if (!c) { assert false; }`, which will reduce the checking of assertion violations to checking whether a specific program line can be reached.

[4]Choosing 32 bit width here is just an arbitrary decision for this example.

```
1   int n = ?(int);                11      while (true) {
2                                   12          wait_time 1;
3   thread generate {              13          assert (n >= 0 && n < 3);
4       while (true) {             14      }
5           n = (n + 1) % 3;       15  }
6           wait_time 1;           16
7       }                          17  main {
8   }                              18      start;
9                                  19  }
10  thread check {
```

Listing 5.2: Example program that demonstrates the usefulness of abstracting away the simulation time when comparing states.

$type(v_1) = type(a_1) = type(a_2)$ and $type(v_2) = type(b_1) = type(b_2)$. Now the $\neg ESS_{\preccurlyeq}$ formula can be formulated as:

$$[\exists x_1, x_2 : (x_1 \neq 0) \wedge (2 * x_1 = v_1) \wedge (x_2 = v_2)] \wedge$$
$$\neg [\exists y_1, y_2, y_3 : (T) \wedge (y_1 + 1 = v_1) \wedge (y_2 + y_3 = v_2)]$$

and passed to the (Z3) SMT solver to check for satisfiability. This formula is unsatisfiable, thus s_1 is subsumed by s_2, denoted as $s_1 \preccurlyeq s_2$.

5.4.3. Relaxing Time Equality Requirements

This section shows that states can be equivalent, even though their simulation times are different. Consider the program in Listing 5.2. The simulation will run indefinitely as no time limit has been specified (the main function just calls start without an optional time limit argument). The program is safe but has actually an infinite state space. The problem is that the simulation time, which is part of the execution state, will be increased in each iteration. But actually the simulation time is not relevant in this example program since the simulation time is not limited and the current time is not accessed from the program directly (and pending notifications store relative time deltas instead of the absolute time to decide when to trigger). Therefore the current simulation time cannot have any influence on the overall simulation result. So it is possible to abstract the current time. Doing so leads to a finite relevant state space (see Figure 5.3) which can be completely explored with stateful model checking.

Listing 5.3 shows an example program, which demonstrates that it is not possible (in general) to abstract away the simulation time, if the simulation is time bounded. It assumes that the @*time* expression returns the current simulation time. Basically the context switch in the while loop will be reached multiple times. All these states are equal except for their simulation time. If the simulation time would not be considered during state matching, then a cycle of states would be detected and the execution path would be stopped. Consequently the assertion violation in Line 5 would falsely not be reached. Another example that does not access the current simulation time directly, but controls the flow of the program indirectly by using the wait_time statement is shown in Listing 5.4. Consider the states s_1 and s_2 reached by $s_0 \xrightarrow{A,B} s_1$ and $s_0 \xrightarrow{B,A} s_2$ from the initial state s_0. They are completely equivalent, except $time(s_1) = 4$ and $time(s_2) = 8$. Thus the notification e will be lost in s_2 since its activation time is beyond the simulation time limit. Now assuming that s_2 is explored first and s_1 is not continued since it is detected as equivalent to s_2, then the trace $s_0 \xrightarrow{C,A,B,C} s_e$ that leads to an error state s_e will be missed.

The general idea used here, to abstract away the simulation time when matching states, can also be naturally extended to other state parts. If a part of the state can be proven to have abso-

```
1  thread A {                          6  }
2    while (@time < 3) {               7
3      wait_time 1;                    8  main {
4    }                                 9    start 5;
5    assert false;                    10  }
```

Listing 5.3: Example program that explicitly retrieves the current simulation time.

```
1  event e;                9                    17     wait e;
2  int n = 1;             10  thread B {        18     assert false;
3                         11    if (n != 0) {   19  }
4  thread A {             12      wait_time 4;  20
5    n = 0;               13    }               21  main {
6    wait_time 4;         14  }                 22    start 9;
7    notify e, 3;         15                    23  }
8  }                      16  thread C {
```

Listing 5.4: Example program that indirectly controls the program flow using wait_time statement.

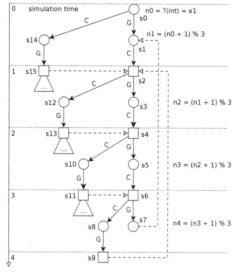

Figure 5.3.: Complete relevant (simulation time abstracted) state space for the program in Listing 5.2. The states are numbered in the order a DFS search could explore the state space. For each delta cycle, either the generator (G) or check (C) thread can be executed first. The resulting state space would be similar. The interleavings are not partial order reduced, since G and C access (with G writing) the same global variable n. But both execution sequences $s \xrightarrow{CG} s'$ and $s \xrightarrow{GC} s'$ result in the same state s' when started in the same state s. Thus (s_{11}, s_6), (s_{13}, s_4) and (s_{15}, s_2) are detected as equivalent states. For this reason neither successor of (s_{11}, s_{13}, s_{15}) has to be explored. Additionally (s_9, s_1) and (s_7, s_2) form cycles. Its necessary to detect them, to prove that the program is indeed safe.

lutely no influence on the simulation result (on the verification of the properties of interest) then it is not necessary to track this part. It turns out to be quite simple to prove for the simulation time (no time limit is specified and the current time is never retrieved from within the program).

6. Heuristic Symbolic Subsumption

A stateful search needs to decide whether a state has already been visited to avoid re-exploration. This process is called *state matching*. The last chapter already presented a general state matching algorithm that works in two steps. First the concrete state parts are matched. And only if they are equal, the symbolic state parts will be checked for subsumption. An exact method, called *Exact Symbolic Subsumption* (ESS), has been presented to detect subsumption between symbolic state parts. The ESS method is exact but computationally very expensive to calculate, especially due to the use of quantifiers. The reason for this complexity is that symbolic expressions can be represented in many (actually infinitely) different ways, e.g. the expressions $a + a$ and $2 * a$ are semantically equivalent but structurally different. Furthermore, a symbolic execution engine manages a path condition, that represents constraints the symbolic values have to satisfy, thus effectively limiting the possible values a symbolic expression can evaluate to.

This chapter presents different heuristic methods to detect subsumption between symbolic state parts more efficiently. They can be used instead of the ESS algorithm as part of the general state matching algorithm, presented in Section 5.4. The goal is to improve the scalability of the overall complete symbolic simulation, by balancing between state matching precision and runtime overhead. It is based on the observation, that spending too much time with state matching can slow down the overall simulation. Often it can be faster to re-explore some (equivalent) part of the state space. Essentially two different heuristic approaches will be presented:

- *Explicit Structural Matching* (ESM) attempts to structurally match all corresponding expression pairs. It offers polynomial worst case complexity but its precision depends on how well the expressions have been normalized in advance.

- *Solver-Based Comparison* (SBC) employs an SMT solver to achieve better precision at the cost of exponential worst case complexity. However, it still offers less runtime overhead than ESS, since no quantifiers are used.

Both ESM and SBC are available in different configurations. The following descriptions assume that the corresponding states are already structurally compatible ($s_1 \sim s_2$), since they only focus on the symbolic state parts. Furthermore the definitions with regard to an execution state, that have been presented in the last chapter in Section 5.1.1, are also relevant for this chapter. As already mentioned, subsumption is a generalization of equivalence, thus detecting equivalence is also a form of subsumption detection. In the following the terms *state subsumption* and *state coverage* will often be used interchangeably.

The rest of this chapter is organized as follows: First Section 6.1 and Section 6.2 present the basic ESM and SBC methods respectively. Section 6.3 will introduce the *fresh symbolic literal problem*. It describes a common situation, that arises when detecting subsumption without using quantifiers, and cannot be handled with the basic ESM and SBC methods. Thus both methods will subsequently be extended in Section 6.4 and Section 6.5 respectively, to solve the *fresh symbolic literal problem*. Thereafter Section 6.6 classifies the ESM and SBC methods together with their extensions by their state matching precision and provides some additional

Algorithm 14: Basic ESM algorithm

Input: Two states s_1 and s_2 with $s_1 \sim s_2$
Output: True iff s_1 is equivalent to s_2

1 **function** $ESM(s_1,s_2)$ **is**
2   **if** $pc(s_1) \neq pc(s_2)$ **then**
3        $\lfloor$ **return** *False*
4     $\lfloor$ **return** $\forall v_1, v_2 \in$ symbolicPairs$(s_1,s_2) : v_1 = v_2$

---

remarks. The ESS algorithm of the previous chapter is also considered in the classification. Finally Section 6.7 sketches some useful improvements for future work.

6.1. Explicit Structural Matching

This section describes a simple method to check, whether two symbolic state parts $\zeta(s_1)$ and $\zeta(s_2)$ are equivalent. The assumption is that s_1 and s_2 are already structurally compatible ($s_1 \sim s_2$). The method is sound (if two states are detected as equivalent they really are - no false positives) but imprecise (some equivalent states might not be detected as such - possible false negative).

First the base method is described in the following section. Then an extension will be presented that increases the detection rate, thus reducing the number of false positives. A further performance optimization will be discussed in Section 6.1.3.

6.1.1. Basic Method

The first heuristic algorithm is based on the simple observation that two (symbolic) expressions must be semantically equal, if they are structurally (syntactically) equal. All linked symbolic values and the path conditions are structurally matched against each other. If they are all completely equal, then both symbolic state parts are equivalent, else they are considered as non equivalent. The principle is shown in Algorithm 14.

This simple algorithm is sufficient to detect all equivalent states of the *simple-counter* example in Listing 6.1, since all of them are structurally completely equal. For example $\zeta(s_6)=(pc: T, c: (x_1 + 1) + 1) = \zeta(s_{19})$, or $\zeta(s_{10})=(pc: T, c: x_1) = \zeta(s_1)$, and so on. The path condition is unconstrained (just *true*) in all states, because it does not happen, that both branches of the (only) conditional statement (Line 15) are feasible.

The advantage of this method to detect necessary equality between symbolic state parts is its polynomial worst case complexity (linear to the node size of all compared expressions). Its disadvantage is however the low precision, since it does not detect semantically equivalent but syntactically different symbolic expressions, except they have been normalized in advance. Improving the algorithm for expression normalization will also directly improve the detection rate of this structural matching. The next section describes some expression simplification rules, that can lead to normalized expressions.

6.1.2. Simplifying Symbolic Expressions

This section describes some expression simplification rules that can lead to normalized expressions.

```
 1   int max = 3;                13   thread reset {              24      while (true) {
 2   int s = ?(int);            14      while (true) {            25        wait CHECK;
 3   int c = s;                 15        if ((c-s) >= max) {     26        assert ((c-s) <=
 4                              16          c = s;                          max);
 5   thread count {             17        }                       27        notify NEXT_STEP;
 6     while (true) {           18        notify CHECK, 0;        28      }
 7       c += 1;                19        wait NEXT_STEP;         29   }
 8       notify CHECK, 0;       20      }                         30
 9       wait NEXT_STEP;        21   }                            31   main {
10     }                        22                                32      start;
11   }                          23   thread verify {              33   }
12
```

Listing 6.1: Symbolic counter example program.

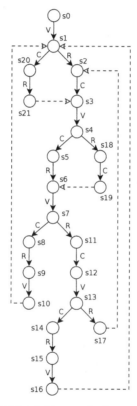

Figure 6.1.: Complete partial order reduced state space for the program in Listing 6.1 for a maximum count of three ($max = 3$). Due to POR it is only necessary to consider interleavings of the threads C (count) and R (reset), since V (verify) is independent to both of them (in the initial state s_0 and later they are no longer co-enabled). This state space could be generated, if a DFS executes the states according to their numbers starting with s_0.

Folding Concrete Arguments

During the simulation, all operations that only involve concrete arguments are directly folded into a single concrete literal (bitvector or logic), e.g. (1 + 2) is directly evaluated to 3. Often operations involve both concrete and symbolic arguments, e.g. $((x_1 + 1) + 1)$ as in state s_6 and s_{19} reached during simulation of Listing 6.1. Clearly this expression can be simplified to (x + 2). A specialized rewrite rule, e.g. $((x+a)+b) \mapsto (x+(a+b))$, can be added, where x will match with any symbolic literal and a, b will match with any concrete literal[1].

Other similar expressions would benefit from the same simplification approach, e.g. ((1 + x) + 1) or (2 + (3 * (x + 2))) and so on. It is possible to add specialized simplification rules for all of them, but there are infinitely many possibilities. But the general idea is always the same. Concrete arguments are reorganized and combined if possible. The idea is to separate the concrete parts from the symbolic parts. The following rules allow to pull concrete expressions to the right side, so that they can be folded. The assumption is that expressions are internally held in binary form (the general idea is also applicable to argument lists). In the following $\oplus$ will denote an commutative operator and $\odot$ an associative operator. The usual associative, commutative and distributive properties also apply to bitvector (integer modulo) arithmetic.

$$a \oplus x \mapsto x \oplus a \tag{6.1}$$
$$(x1 \oplus a) \oplus x2 \mapsto (x1 \oplus x2) \oplus a \tag{6.2}$$
$$(x1 \oplus a) \oplus (x2 \oplus b) \mapsto (x1 \oplus x2) \oplus (a \oplus b) \tag{6.3}$$
$$(x1 \oplus a) \oplus (x2 \oplus x3) \mapsto ((x1 \oplus x2) \oplus x3) \oplus a \tag{6.4}$$
$$(x \odot a) \odot b \mapsto (x \odot (a \odot b)) \tag{6.5}$$
$$x1 \odot (x2 \odot a) \mapsto (x1 \odot x2) \odot a \tag{6.6}$$
$$(x1 \odot x2) \odot (x3 \odot a) \mapsto ((x1 \odot x2) \odot x3) \odot a \tag{6.7}$$

Additionally specialized simplification rules can be defined to simplify expressions with mixed operators or perform term expansion or factoring.

$$(x-a)+b \mapsto x+(b-a) \tag{6.8}$$
$$(x+a)-b \mapsto x+(a-b) \tag{6.9}$$
$$(x*a)+(x*b) \mapsto x*(a+b) \tag{6.10}$$
$$(x+a)*b \mapsto (x*b)+(a*b) \tag{6.11}$$

It is not directly apparent whether more complex terms like $(x1 + a) * (x2 + b) \equiv (((x1 * x2) + (x1 * b)) + (x2 * a)) + (a * b)$ should be expanded or factored, to achieve a better normalization. Depending on the use case (the actual program that shall be verified) it might be useful to add some additional specialized simplification rules.

Using the above rules, expressions like $e = (1 + x_1) + (2 * ((x_2 + 7) - 4))$ will no longer occur, since they would already have been simplified. All expressions are simplified on construction. Literal expressions are already fully simplified by definition. This implies, that every compound expression constructor will receive argument expressions, which already are fully simplified (with respect to the simplification rules). This allows to apply the rules locally and yield a fully simplified expression, without a complete bottom up expression tree transformation. Thus when attempting to construct e, the following (sub) expressions will be constructed and simplified

---

[1] A previous semantic analysis (or a similar mechanism) ensures that the operands are already type compatible.

```
 1  int pressure = ?(int);      13  }                          25      wait_time 1;
 2  int max = 9;                14                             26      notify e, 0;
 3  int bound = 4;              15  thread increase {          27  }
 4  event e;                    16    while (true) {           28  }
 5                              17      wait e;                29
 6  thread guard {              18      pressure += 1;         30  main {
 7    while (true) {            19    }                        31    assume pressure <=
 8      wait e;                 20  }                                bound && pressure
 9      if (pressure >= max)    21                                   >= 0;
       {                        22  thread clk {               32    start 9;
10      pressure -= 1;          23    while (true) {           33  }
11    }                         24      assert pressure <=
12  }                                   max;
```

Listing 6.2: Example program that demonstrates the usefulness of symbolic simplification rules that fold concrete arguments.

(construction is performed bottom up, starting at the leafs and proceeding to the root):

$$1 + x_1 \mapsto x_1 + 1 \qquad \text{due to 6.1}$$
$$x_2 + 7 \mapsto x_2 + 7$$
$$(x_2 + 7) - 4 \mapsto x_2 + 3 \qquad \text{due to 6.9}$$
$$2 * (x_2 + 3) \mapsto (x_2 + 3) * 2 \qquad \text{due to 6.1}$$
$$(x_2 + 3) * 2 \mapsto (x_2 * 2) + 6 \qquad \text{due to 6.11}$$
$$(x_1 + 1) + ((x_2 * 2) + 6) \mapsto (x_1 + (x_2 * 2)) + 7 \qquad \text{due to 6.3}$$

Yielding the expression $(x_1 + (x_2 * 2)) + 7$ as the result.

The symbolic version of the *pressure* program[2], shown in Listing 6.2, is an example program, where these simplifications are useful. It is a safe program that runs for a specified number of time steps. Initially the variable *pressure* is assigned a symbolic integer value designated as x_1 and constrained to be in the range [0..bound]. During the simulation, the threads *increase* and *guard* will increase or respectively decrease, once a maximum value is reached, the current value of *pressure*. So the value of the *pressure* variable will in principle have the form: $(((((((x_1 + 1) + 1) + 1) - 1) + 1) - 1)...)$, it first starts increasing and then alternately swap between increase and decrease until the time bound is reached. Without simplification rules, the structural matching algorithm will miss all equivalent states, since neither value of pressure will be structurally equal. With (the above) simplification rules the value of *pressure* will always stay in the form $(x_1 + i)$, where $i \in \{0, ..., max + 1\}$[3].

Eliminating Redundant Operations

The binary subtraction operator is neither commutative nor associative and thus the above simplification rules 6.1-6.7 could not be applied. Again specialized rules like $(x - a) + b \mapsto x + (b - a)$ can be used to simplify $((x - 1) + 2)$, but similar expressions like $((1 - x) + 2)$ will require additional rules. Instead subtraction can be rewritten by using addition and unary subtraction as $x - y \mapsto x + (-y)$. Such a reduction can occur in a preprocessing phase before the actual simulation, e.g. right after parsing and before the semantic analysis. By eliminating binary subtraction completely, it is no longer necessary to consider it during simplification. Existing commutative and associative simplification rules can now be applied to simplify e.g.

---

[2]It is based on a version that appeared in [BK10]. However it has been adapted to be safe, by decreasing the simulation time accordingly.

[3]Due to scheduling interleavings, *pressure* can temporarily reach the value $max + 1$ in the thread *increase*.

```
1   bool b = true;              17   while (true) {            33   for (step=0;
2   int n = ?(int);            18      wait e;                       step<15;
3   int step = 0;              19      extend_cond(step);            step+=1) {
4   bool is_first = true;      20   }                        34      is_first = true;
5   event e;                   21   }                        35      notify e;
6                              22                             36      wait_time 0;
7   void extend_cond(int i) {  23   thread B {               37   }
8     if (i < 7 || i > 8) {    24      while (true) {        38   assert (!b || (n ==
9       if (!is_first)         25         wait e;                      7 || n == 8));
10         assert (!b || (n    26         extend_cond(15-step);  39   }
              != (15-i)));     27      }                      40   }
11         is_first = false;   28   }                         41
12         b = b && (n != i);  29                             42   main {
13     }                       30   thread C {               43      assume (n >= 0 && n <=
14   }                         31      wait_time 0;                   15);
15                             32      while (true) {         44      start;
16   thread A {                                               45   }
```

Listing 6.3: *condition-builder*, example program whose simulation would produce many equivalent states, due to commutative operations.

$((1-x)+2)$ as $((1-x)+2) \mapsto (((-x)+1)+2) \mapsto ((-x)+3)$.

$$a - b \mapsto a + (-b) \tag{6.12}$$

$$a \geq b \mapsto (a > b) \vee (a = b) \tag{6.13}$$

$$a \neq b \mapsto \neg(a = b) \tag{6.14}$$

$$a \leq b \mapsto (a < b) \vee (a = b) \tag{6.15}$$

$$a > b \mapsto \neg(a < b) \wedge \neg(a = b) \tag{6.16}$$

The main advantage of this normalization, is that it reduces the number of possibilities to create semantically equal expressions with syntactically different constructs. This leads to better normalized expressions and thus a higher detection rate of the structural matching algorithm. It also reduces the number of operators that need to be implemented. A slight disadvantage of this approach is, that it may (minimally) reduce the simulation performance. The reason is that a single operator expression is rewritten with multiple operators, e.g. $(a > b)$ is rewritten to an expression that contains five operators.

Simplifying Special Cases

Applying the same unary operator twice cancels out their effects, thus $\circ(\circ x) \mapsto x$ with $\{-, !, \sim\} \in \circ$. Other operations that either have no effect or produce a constant literal are shown in the following.

$$x + 0 \mapsto x \qquad x * 0 \mapsto 0 \qquad x * 1 \mapsto x \qquad x | 0 \mapsto x$$
$$x | x \mapsto x \qquad x \& x \mapsto x \qquad x \wedge T \mapsto x \qquad x \wedge F \mapsto F$$
$$x \vee T \mapsto T \qquad x \vee F \mapsto x \qquad x \wedge x \mapsto x \qquad x \vee x \mapsto x$$
$$x \wedge \neg x \mapsto F \qquad x \vee \neg x \mapsto T$$

It is sufficient to specify this rules once with the concrete operand being on the right side, since it will be transfered to the right side as already described. The operators | and & denote bitwise *or* respectively *and* operations.

The direct absorption rules $x \wedge x \mapsto x$ and $x \vee x \mapsto x$ can be extended to work with a list of arguments $(x_1...x_n)$.

$$\left(\bigwedge_{i\in\{1...n\}} x_i \right) \wedge x \mapsto \left(\bigwedge_{i\in\{1...j-1,j+j...n\}} x_i \right) \text{ if } x_j = x \tag{6.17}$$

$$\left(\bigvee_{i\in\{1...n\}} x_i \right) \vee x \mapsto \left(\bigvee_{i\in\{1...j-1,j+j...n\}} x_i \right) \text{ if } x_j = x \tag{6.18}$$

Doing so will remove any (syntactically) duplicate expressions from logic expressions. The bitwise *and* and *or* operations can be rewritten analogously.

Adding the same term in negated form to a conjunction (or disjunction) list, results in a contradiction (or tautology). The direct simplification rules are $x \wedge \neg x \mapsto F$ and $x \vee \neg x \mapsto T$. Similarly to the absorption rules they can also be extended to work with a list of arguments $(x_1...x_n)$.

$$\left(\bigwedge_{i\in\{1...n\}} x_i \right) \wedge x \mapsto F \text{ if } x = \neg x_j \tag{6.19}$$

$$\left(\bigvee_{i\in\{1...n\}} x_i \right) \wedge x \mapsto T \text{ if } x = \neg x_j \tag{6.20}$$

Further Improvement

Another promising optimization, which has not yet been implemented, is to sort all expression arguments based on some total order between expressions. This allows to normalize all commutative operations, that have the same arguments but possibly in different order. An example program that demonstrates the usefulness of such an approach is available in Listing 6.3.

The program is safe, it runs indefinitely but has a finite state space, since every possible execution path runs into a cycle. In this case, the program will repeat its behaviour after $k = 16$ steps.

Essentially the program manages a boolean condition $b = True$ and a symbolic integer $n = x_1$, which is initially constrained to be in the range $[0..k-1]$. The thread C repeats the following behaviour indefinitely: it runs for k steps and then verifies the constructed condition. In each step C notifies the threads A and B. The threads A and B will update the condition b in each step i (starting with 0, ending with $k-1$), where $i \neq \lceil (k-1)/2 \rceil$ and $i \neq \lfloor (k-1)/2 \rfloor$, to either (A) $b \wedge (n \neq i)$ or (B) $b \wedge (n \neq (k-1-i))$. So b_{i+1} will have either the new value $b_i \wedge (n \neq i) \wedge (n \neq (k-1-i))$ or $b_i \wedge (n \neq (k-1-i)) \wedge (n \neq i)$, depending whether $\xrightarrow{A,B}$ or $\xrightarrow{B,A}$ is executed.

Regardless of the scheduling decision, the same terms will be added to b, but they appear in different orders. All terms can be arbitrarily rearranged since the logic *and* operator is commutative. The resulting expression (after sorting the arguments) depends on the total order. For this use case, the choice of the total order is actually completely irrelevant. Any arbitrary valid order will suffice.

But even without detecting commutative operations, the program can still be proven correct, because the number of different terms added to b is finite. Each term has the form $n \neq j$ with $j \in (\{0..\lfloor (k-1)/2 \rfloor - 1\} \cup \{0..\lceil (k-1)/2 \rceil + 1\})$. Thus using the absorption simplification rule 6.17 a fixpoint is reached after two iterations for each execution path (since each iteration adds the same terms). It *only* takes (considerably) longer to explore the relevant state space completely, since the execution paths that arise due to different scheduling decisions are not pruned earlier.

6.1.3. Hashing of Symbolic State Parts

The performance of the structural matching algorithm can be (significantly) improved, if one can detect early that two expressions must have a structural incompatibility in which case they cannot be matched.

A common practice to filter out necessary unequal objects early before the actual comparison is to compute a hash value. The generic comparison Algorithm 13 does not do so, because symbolic values can have many different syntactic representation with the same semantic. But since the structural matching algorithm will only match completely equal expressions, it is easy to define a hash function that preserves the property: $hash(e_1) \neq hash(e_2) \implies e_1 \neq e_2$ for all expressions e_1 and e_2, where $e_1 \neq e_2$ returns $True$ if e_1 and e_2 would not match using the ESM algorithm.

Thus two symbolic state parts $\zeta(s_1)$ and $\zeta(s_2)$ can only be equivalent, if all value pairs $(v_1, v_2) \in \zeta(s_1, s_2)$ and both path conditions have the same hash value. Therefore the hash value of the complete state can also include the hash values of the symbolic state part. This allows the generic comparison Algorithm 13 presented in the overview Section 5.4.1 to filter out necessarily non-equivalent states (states that will not be detected as equivalent with the structural match method) early. The hash value can be computed once during the construction of each expression effectively in constant time, since it has already been computed for every child node.

The rational behind this optimization is based on the observation that two symbolic expressions most often are not structurally identical. Thus it is especially useful, if many structurally compatible states are generated (which means that symbolic state parts will be compared often) and symbolic expressions grow quite large and do not directly mismatch.

6.2. Solver-Based Comparison

The Explicit Structural Matching (ESM) algorithm, described in the previous section, strongly depends on a good expression normalization. It is a non-trivial task to devise and implement algorithms that normalize expressions extensively and efficiently. Instead it would be more useful to leverage available SMT solvers to decide whether some (arbitrary) expressions are semantically equivalent. The following description starts with a first algorithm, that reduces the equivalence checking problem of symbolic state parts to the satisfiability of an SMT formula. Similarly to the ESM algorithm, it will be assumed that the compared states are already structurally compatible. So their concrete state parts are completely equal. Only their symbolic state parts will be compared.

6.2.1. Detecting Equivalent States

This section describes, how the equality checking of the symbolic parts of two execution states s_1 and s_2 can be reduced upon the satisfiability of an SMT formula. To this end an SMT formula is generated, that is satisfiable, if the state parts can be unequal. It implies, that if the formula is unsatisfiable, then the state parts must be equivalent.

```
1   event COPY;                      31                                  60   int num_high_bits(u32 x)
2   u32 c = ?(u32);                  32   thread copy_1 {                      {
3   u32 n = 0;                       33     while (true) {                61     int count = 0;
4                                    34       rotate_byte(&n, c,         62     for (int i=0; i<32; i
5   void rotate_byte(u32             0);                                         += 1) {
      *dst, u32 src, int   35        wait COPY;                 63       count += (x & (1 <<
      pos) {              36       }                                              i)) >> i;
6     u32 mask;                      37   }                             64     }
7                                    38                                  65     return count;
8     switch (pos) {                 39   thread copy_2 {                66   }
9     case 0:                        40     while (true) {
10      mask = 0x000000ff;           41       rotate_byte(&n, c,        68   void check_state() {
11      break;                                  1);                      69     int a =
12    case 1:                        42       wait COPY;                          num_high_bits(c);
13      mask = 0x0000ff00;           43     }                           70     int b =
14      break;                       44   }                                       num_high_bits(n);
15    case 2:                        45                                  71     assert a == b;
16      mask = 0x00ff0000;           46   thread copy_3 {                72   }
17      break;                       47     while (true) {
18    case 3:                        48       rotate_byte(&n, c,        74   thread clk {
19      mask = 0xff000000;                      2);                      75     while (true) {
20      break;                       49       wait COPY;                 76       wait_time(1);
21    default:                       50     }                           77       notify COPY;
22      assert false;                51   }                             78
23    }                              52                                  79       check_state();
24                                   53   thread copy_4 {                80       c = n;
25    if (pos == 3) {                54     while (true) {               81       n = 0;
26      *dst |= ((src &              55       rotate_byte(&n, c,        82     }
        mask) >> 24);                          3);                       83   }
27    } else {                       56       wait COPY;                 84
28      *dst |= ((src &              57     }                           85   main {
        mask) << 8);                 58   }                             86     start 2;
29    }                              59                                  87   }
30  }
```

Listing 6.4: *rbuf*, example program that demonstrates the benefits of a symbolic state comparison method that is insensitive to the argument order of commutative operations.

Definition 21

Two symbolic state parts s_1 and s_2 are equivalent, i.e. $s_1 \simeq s_2$ according to Definition 15, if $F_\simeq(s_1, s_2)$ is valid.

$$F_\simeq(s_1, s_2) = (pc(s_1) = pc(s_2)) \wedge (\bigwedge_{(v_1, v_2) \in symbolicPairs(s_1, s_2)} v_1 = v_2)$$

Similarly to the ESS method, as presented in Section 5.4.2, the formula $F_\simeq(s_1, s_2)$ is valid, if $\neg F_\simeq(s_1, s_2)$ is unsatisfiable.

$$\neg F_\simeq(s_1, s_2) = (pc(s_1) \neq pc(s_2)) \vee (\bigvee_{(v_1, v_2) \in symbolicPairs(s_1, s_2)} v_1 \neq v_2)$$

An SMT solver can be queried in order to check the above (negated) formula for satisfiability. The advantage of this state equivalence matching method is, as already stated, that it does not require to normalize expressions in order to detect equivalences. Also it is insensitive to the argument order of symmetric operations. For this reason it performs very well for e.g. the example program in Listing 6.3.Since the current normalization method described in Section 6.1.2 does not sort expression arguments based on some total order, it misses many equivalent states which the solver-based method does detect. Another example program which demonstrates this observation even more clearly is shown in the following.

Remark. Normalization of expressions is still an important part of a symbolic execution engine, since it can reduce the size of expressions significantly, which directly benefits the performance of the underlying SMT solver.

Example: Rotating (Bit-)Buffer

This example program demonstrates the benefits of a symbolic state comparison method that is insensitive to the argument order of commutative operations. The solver-based method is a representative example for that. The program is shown in Listing 6.4. In the following it will often be referred to as *rbuf*. It is safe and runs for a fixed number of time steps. Two global 32 bit values are managed. A current value c and a next value n. In each time step the next value is computed from the current value as follows:

$$n[0..7]\ \ = c[24..31]$$
$$n[8..15]\ = c[0..7]$$
$$n[16..23] = c[8..15]$$
$$n[24..31] = c[16..23]$$

So bit 0 of n will receive the value of bit 24 of c, and so on. The four threads $copy_1..copy_4$ are responsible to copy the four bytes of c as described above. The clock thread clk updates the current value, resets the next value to 0 and notifies all copy threads. This starts the next time step.

All four copy threads are dependent, since they write to the same global variable. Thus all their $4! = 24$ interleavings have to be considered in each time step. If none of them can be detected to be equivalent, then each execution path will generate 24 new paths in each time step. After n time steps there would be a total number of 24^n execution paths. But using the solver-based symbolic state comparison method, all interleavings of a time step will be detected to be equivalent. Thus only a single execution path remains after each time step.

6.2.2. Detecting State Subsumption

The Definition 21 can be adapted to detect subsumed/covered instead of equivalent state parts.

Definition 22

> A symbolic state part s_1 is covered/subsumed by s_2, i.e. $s_1 \preccurlyeq s_2$ according to Definition 14, if $F_{\preccurlyeq}(s_1, s_2)$ is valid.
>
> $$F_{\preccurlyeq}(s_1, s_2) = \left(pc(s_1) \implies \left(pc(s_2) \land \bigwedge_{(v_1, v_2) \in \texttt{symbolicPairs}(s_1, s_2)} v_1 = v_2 \right) \right)$$

Again the formula $F_{\preccurlyeq}(s_1, s_2)$ is valid, if its negation $\neg F_{\preccurlyeq}(s_1, s_2)$ is unsatisfiable.

$$\neg F_{\preccurlyeq}(s_1, s_2) = \left(pc(s_1) \land \left(\neg pc(s_2) \lor \bigvee_{(v_1, v_2) \in \texttt{symbolicPairs}(s_1, s_2)} v_1 \neq v_2 \right) \right)$$

By definition it immediately follows that $F_{\simeq}(s_1, s_2)$ implies $F_{\preccurlyeq}(s_1, s_2)$ and $F_{\preccurlyeq}(s_2, s_1)$. But the reverse is not true. Consider for example the symbolic state parts $s_1 = (pc\colon x_1 > 5,\ a\colon x_1 > 5\ ?\ 4 : 1)$ and $s_2 = (pc\colon x_1 > 5,\ a\colon x_1 > 5\ ?\ 4 : 2)$. They would not be detected equivalent by Definition 21 of $F_{\simeq}(s_1, s_2)$, but both $F_{\preccurlyeq}(s_1, s_2)$ and $F_{\preccurlyeq}(s_2, s_1)$ are valid. Similarly for the symbolic state parts $s_1=(pc: (x_1 > 9), a: x_1)$ and $s_2=(pc: (x_1 > 5), a: x_1)$, $F_{\preccurlyeq}(s_1,s_2)$ is valid, whereas $F_{\approx}(s_1,s_2)$ is not. Thus Definition 22 of $F_{\preccurlyeq}(s_1,s_2)$ is a strictly weaker state matching criteria than Definition 21 of $F_{\approx}(s_1,s_2)$. A complete example, where this heuristic coverage/subsumption method would perform considerably better, than requiring equivalent states, is presented in the following.

Example: (Bounded) Symbolic Counter

This example demonstrates the benefits of detecting covered states using Definition 22 instead of detecting equivalent states using Definition 21. The former will be able to handle this example with larger parameters whereas the latter approach will not scale.

The program is available in Listing 6.5. It manages a single symbolic counter c, hence the program will be referred to as *symbolic-counter*. Initially c is assigned a symbolic value denoted as x_1 (Line 4), which is constrained in Line 32 to $x_1 \in \{0,...,bound\}$ (in this example *bound* = 1). The thread *count* will increment the counter in each step. Another thread *guard* will reset the counter to its initial range once a configurable maximum value *max* is reached (in this example *max* = 4). The interleavings of these threads are not reduced by POR, since they both access the counter (with at least one write access). But all of them, except those where the *guard* thread resets the counter, can be reduced by a stateful space exploration, since the *guard* thread, in all other cases, does not change the execution state at all. The thread *check* just validates that the counter is in the range $[0..max]$ and notifies the other threads, so they become runnable again. The symbolic part of the initial state is $\zeta(s_0) = (pc: I, c: x_1)$, where $I = x_1 \geq 0 \wedge x_1 \leq 1$. Until the counter does not exceed the maximum value, it will normally be incremented in each step. Thus the symbolic state parts $\zeta(s_1)=(pc: I, c: x_1 + 1)$ and $\zeta(s_2)=(pc: I, c: x_1 + 2)$ will be produced. Whenever the counter c can exceed the maximum value *max*, the *guard* thread will reset it to $c = c - max + bound$. So c will be in the range $[0..bound + 1]$ again. Since c is symbolic both paths are feasible at the (first) check $c \geq max$, the execution path is forked into s_T and s_F, and their path conditions are extended correspondingly as $pc(s_T)\wedge = c \geq max$ and $pc(s_F)\wedge = c < max$. So the counter of s_T is reset, whereas the counter of s_F will be reset the next time *guard* is called after increasing the counter. Or s_F will analogously fork itself. The number of non-equivalent forked execution paths corresponds with the number of possible initial values of the counter c (which is $x_1 \in \{0,...,bound\}$). In this case there will be two non-equivalent forked execution paths that will be reset, their symbolic state parts are: $s_{r1}=(pc: I \wedge (x_1 + 2 \geq max), c: x_1) \preccurlyeq s_0$ and $s_{r2}=(pc: I \wedge (x_1 + 2 < max), c: x_1 + 1) \preccurlyeq s_1$.

These states are covered by s_0 (and s_1) because the negations of $F_{\preccurlyeq}(s_{r1},s_0)$ and $F_{\preccurlyeq}(s_{r2},s_1)$ respectively, are unsatisfiable. So basically s_{r1} and s_{r2} have already been explored with less constraints. But they are not equivalent with any previously explored state, because the path conditions have been extended. Thus they would need to be re-executed until their counter hits the maximum value again. The condition $c \geq max$ will then only have one feasible path (due to the extended path condition) so the counter will be reset again, but the path condition will not be extended. So the state will be detected to be equivalent with either s_{r1} or s_{r2}.

6.3. Fresh Symbolic Literal Problem

The previous example programs used as motivation for different state matching methods (shown in Listing 6.1, Listing 6.2, Listing 6.3, Listing 6.4 and Listing 6.5) create all symbolic values once at the beginning of the simulation. They were shared among all execution paths and thus could be matched against each other. A common behaviour is to create fresh symbolic values

```
1   int MAX = 3;              14   thread guard {            26        assert (c >= 0 && c
2   int BOUND = 1;            15     while (true) {                         <= MAX);
3                             16       if (c >= MAX) {       27        notify e;
4   int c1 = ?(int);          17         c = c - MAX +       28      }
5   event e;                  18             BOUND;          29   }
6                             18       }                     30
7   thread counter {          19       wait e;               31   main {
8     while (true) {          20     }                       32      assume (c1 >= 0 && c1
9       c += 1;               21   }                                     <= BOUND);
10      wait e;               22                             33      start;
11    }                       23   thread check {            34   }
12  }                         24     while (true) {
13                            25       wait_time 0;
```

Listing 6.5: *symbolic-counter*, example program that demonstrates the advantages of a stateful exploration that detects covered states instead of equivalent ones.

during the simulation. The *tokenring* program, shown in Listing 6.6, is a simple example that demonstrates this behaviour.

The program is safe. Its complete state space is shown in Figure 6.2. It consists of two threads *master* (M) and *transmit* (T), that communicate using a global variable *token* and two events E1 and EM. Initially both threads are enabled. If the simulation starts with the master thread, then the notification E1 of the master thread will be lost, thus the execution path terminates after executing the transmit thread. This case corresponds to the trace $s_0 \xrightarrow{M,T} s_2$ (in Figure 6.2).

Choosing transmit first and then master results in the state s_4. From here on, both threads will alternately indefinitely. But they repeat the same behaviour in cycles. Thus the state space is finite.

Every time the master thread is executed, it generates a fresh symbolic literal and assigns it to the token variable. A copy is also stored locally. Before yielding control to the simulation kernel, it enables the transmit thread. The transmit thread then increments the token variable and notifies the master thread. The master thread then verifies the value of the token variable and the cycle starts again.

The states s_4 and s_6 are equivalent. Their symbolic state parts are $\zeta(s_4) = (pc = T, token = x_1, local = x_1)$ and $\zeta(s_6) = (pc = T, token = x_2, local = x_2)$. But they would not be detected as such by the algorithms presented so far. The reason is that x_1 and x_2 are different symbolic literals. Thus they would not be matched by the ESM method. But actually they should be, since they are both unconstrained and have the same type. Similarly the SBC method will fail, e.g. using the Definition 22 (using Definition 21 would fail similarly) to construct a formula for state subsumption checking results in:

$$T \implies (\neg T \vee (x_1 \neq x_2))$$

This formula clearly is satisfiable, e.g. $\psi = \{x_1 = 0, x_2 = 1\}$ would be a model, which means the states will not be matched. The idea is to strengthen the formula with additional equality assumptions between symbolic literals.

The *fresh symbolic literals problem* only arises in the context of heuristic methods, the ESS algorithm does not have this problem, because every symbolic literal is bound by a quantifier. The following two sections present extended versions of the ESM and SBC methods that can handle fresh symbolic literals.

```
 1  event E1;              11      notify E1;         20      token = token + 1;
 2  event EM;              12      wait EM;           21      notify EM;
 3  int token;             13      assert token ==    22    }
 4                         14                (local + 1);   23  }
 5  thread master {        14    }                    24
 6    int local;           15  }                      25  main {
 7                         16                          26    start;
 8    while (true) {       17  thread transmit {      27  }
 9      token = ?(int);    18    while (true) {
10      local = token;     19      wait E1;
```

Listing 6.6: Tokenring example program with one transmitter thread.

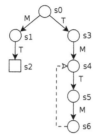

Figure 6.2.: Complete state space of the tokenring example program in Listing 6.6 with one transmitter thread. The dotted line from s_6 to s_4 represents, that these states are equivalent.

6.4. Extended Explicit Structural Matching

This section presents an extended ESM method, that can handle the fresh symbolic literals problem as introduced in the previous section. The idea is that symbolic literals are not required to be equal (identical objects) it is sufficient that they have the same type. Every symbolic literal is allowed to be matched with at most one other literal to prevent false positives (some examples will be shown in the following). Since the path condition is also matched, this ensures, that matched symbolic literals have the same constraints on both states.

6.4.1. Extended Algorithm

An extended algorithm is shown in Algorithm 15. Two states s_1 and s_2 will be considered equivalent, iff the $ESM(s_1,s_2)$ function returns True.

All linked symbolic values and the path conditions are structurally matched against each other as in the base algorithm Section 6.1.1. This matching will fail, if any of those value pairs is not compatible in general (e.g. a binary expression is matched against an unary one, or the operators are not equal, and so on). Symbolic literals are no longer required to be identical objects, it is sufficient if they have the same type. During the matching, a symmetric mapping m between symbolic literal expressions is constructed. Every time a pair of symbolic literals (x,y) is matched it will first be checked, whether this contradicts any existing matched pair (a,b), which happens if $\exists (a,b) \in m : (x = a \wedge y \neq b)$. Else (x,y) will be recorded in m (and also (y,x) since m is symmetric). Two symbolic state parts are equivalent, if all linked symbolic values and the path conditions can be matched without contradiction starting with an empty symbolic literal mapping m. Otherwise they will be considered non-equivalent.

Algorithm 15: Extended ESM algorithm

Input: Two states s_1 and s_2 with $s_1 \sim s_2$
Output: True iff s_1 is equivalent to s_2

1 **function** *ESM(s_1,s_2)* **is**
2     $m \leftarrow \{\}$
3   **if** $\neg$matchExpressions(pc(s_1), pc(s_2), *m*) **then**
4   **return** *False*
5   **return** $\forall v_1, v_2 \in$ symbolicPairs(s_1,s_2) : matchExpressions(v_1, v_2, *m*)

6 **function** *matchExpressions(e_1, e_2, m)* **is**
7   **if** isPrimitive(e_1) $\wedge$ isPrimitive(e_2) **then**
8   **return** matchPrimitives(e_1, e_2, *m*)
9   **else**
10   **return** matchCompounds(e_1, e_2, *m*)

11 **function** *matchPrimitives(e_1, e_2, m)* **is**
12   **if** isConcrete(e_1) $\wedge$ isConcrete(e_2) **then**
13   **return** $e_1 = e_2$
14   **else if** isSymbolic(e_1) $\wedge$ isSymbolic(e_2) **then**
15   **if** type(e_1) $\neq$ type(e_2) **then**
16   **return** *False*
17   **else if** $e_1 \in m$ **then**
18   **if** $m[e_1] \neq e_2$ **then**
19   **return** *False*
20   **else**
21           $m.add((e_1, e_2))$
22           $m.add((e_2, e_1))$
23   **else**
24   **return** *False*
25   **return** *True*

26 **function** *matchCompounds(e_1, e_2, m)* **is**
27   **if** getOperator(e_1) $\neq$ getOperator(e_2) **then**
28   **return** *False*
29   **return** $\forall a1, a2 \in$ zip(getOperands(e_1), getOperands(e_2)) : matchExpressions(a_1, a_2)

6.4.2. Examples

All the following examples assume, that the (matched) symbolic literals have equal types. The states s_4 and s_6 of the *tokenring* example with symbolic parts $\zeta(s_4)$=$(pc\colon T, token\colon x_1, local\colon x_1)$ and $\zeta(s_6)$=$(pc\colon T, token\colon x_2, local\colon x_2)$ can be matched with the mapping m={(x_1, x_2)}. The symbolic state parts $\zeta(s_1)$=$(pc\colon (x_1 < 5) \wedge (x_2 > 4), a\colon 2*(x_1 + x_3), b\colon x_1*(x_2 + 1), c\colon x_1 < x_3)$ and $\zeta(s_2)$=$(pc\colon (x_1 < 5) \wedge (x_4 > 4), a\colon 2*(x_1 + x_5), b\colon x_1*(x_4 + 1), c\colon x_1 < x_5)$ can be matched with m={$(x_1,x_1), (x_2,x_4), (x_3,x_5)$}.

The following example demonstrates why it is necessary to require that every symbolic literal is matched with at most one other symbolic literal. Consider two symbolic state parts $\zeta(s_1)$=$(pc\colon T, v\colon x_1 + 1, w\colon x_2 + 1)$ and $\zeta(s_2)$=$(pc\colon T, v\colon x_1 + 1, w\colon x_1 + 1)$. Starting with an empty mapping m and matching v_1 with v_2 results in $m = \{(x_1,x_1)\}$. Now matching w_1 and w_2 using m results in an inconsistent mapping $m = \{(x_1,x_1), (x_1,x_2)\}$. Thus s_1 and s_2 are correctly classified as non equivalent. Executing the statement `assert(v == w)` on both states independently would pass on s_2 but fail on s_1, e.g. ψ={$x_1 = 0, x_2 = 1$} would be a model that satisfies $v_1 \neq w_1$.

6.4.3. Comparison with Base Version

The advantages and disadvantages of the base algorithm also apply to this extended version. The matching itself has a polynomial runtime complexity but its detection rate is highly dependent on a good expression normalization. This extension is never less precise than the base version. Every state pair that is detected by the base algorithm to be equivalent, is also detected by this extended algorithm. The base version can be reduced upon the extended version, by restricting the valid symbolic literal mappings to those that only allow to map a literal to itself. This extended version can also hash symbolic state parts, as the base version. The hash function has to be slightly adapted though, since symbolic literals are matched differently. Only type informations but not the name (or memory address, depending on the implementation) of a symbolic literal is included in the computation of a hash value. All other expressions can reuse the same hash function. The additional runtime overhead of this algorithm compared to the base version is negligible. Thus the extension is truly superior to the base version.

6.5. Extended Solver-Based Comparison

This section presents an extended solver-based state matching, that can handle the fresh symbolic literals problem as introduced in Section 6.3. To detect equivalence between expressions which include fresh symbolic literals, additional equality assumptions are assumed between them. A set of equality assumptions will also be referred to as symbolic literal mapping m. They form a irreflexive, symmetric binary relation over all symbolic literals. The equality assumptions (symbolic literal mapping) cannot be chosen arbitrarily. Every symbolic literal is allowed to be mapped to at most one other literal. And mapped literals must be type compatible. Equality assumptions that respect these two properties are called consistent. The following three definitions state this intuitive description more formally.

Definition 23

A *symbolic literal mapping m between two states s_1 and s_2 is ambiguous iff there exist a symbolic literal, that is mapped to more than one other symbolic literal, i.e.* $\exists (a_1, b_1), (a_2, b_2) \in m : a_1 = a_2 \wedge b_1 \neq b_2$.

Definition 24

> A symbolic literal mapping m between two states s_1 and s_2 is type-compatible iff all symbolic literals that are mapped on each other have the same type ($\forall(a,b) \in m : type(a) = type(b)$).

Definition 25

> A symbolic literal mapping (set of equality assumptions) m is consistent iff it is type-compatible and not ambiguous. Else it is called inconsistent.

Given a set of consistent equality assumptions, the definitions 21 and 22 of the base algorithm can be extended, in order to solve the fresh symbolic literal problem as discussed in the previous section, as follows.

Definition 26

> Two states s_1 and s_2 are equivalent, i.e. $s_1 \simeq s_2$ according to Definition 15, if there exists a set of consistent equality assumptions m, such that $F_{\simeq}(s_1, s_2, m)$ is valid.
>
> $$F_{\simeq}(s_1, s_2, m) = \left[\left(\bigwedge_{(x,y)\in m} x = y \right) \implies F_{\simeq}(s_1, s_2) \right]$$
>
> The formula $F_{\simeq}(s_1, s_2)$ of Definition 21 is reused here.

Definition 27

> A state s_1 is covered/subsumed by a state s_2, i.e. $s_1 \preccurlyeq s_2$ according to Definition 14, if there exists a set of consistent equality assumptions m, such that $F_{\preccurlyeq}(s_1, s_2, m)$ is valid.
>
> $$F_{\preccurlyeq}(s_1, s_2, m) = \left[\left(\bigwedge_{(x,y)\in m} x = y \right) \implies F_{\preccurlyeq}(s_1, s_2) \right]$$
>
> The formula $F_{\preccurlyeq}(s_1, s_2)$ of Definition 22 is reused here.

Again the formulas $F_{\simeq}(s_1, s_2, m)$ and $F_{\preccurlyeq}(s_1, s_2, m)$ are valid, if their negations $\neg F_{\simeq}(s_1, s_2, m)$ and $\neg F_{\preccurlyeq}(s_1, s_2, m)$ respectively, are unsatisfiable.

$$\neg F_{\simeq}(s_1, s_2, m) = \left[\left(\bigwedge_{(x,y)\in m} x = y \right) \wedge \left(\neg F_{\simeq}(s_1, s_2) \right) \right]$$

$$\neg F_{\preccurlyeq}(s_1, s_2, m) = \left[\left(\bigwedge_{(x,y)\in m} x = y \right) \wedge \left(\neg F_{\preccurlyeq}(s_1, s_2) \right) \right]$$

This extension is a real generalization of the previous solver-based method (which itself can be considered a generalization of the ESM method). Because if an empty symbolic literal mapping $m = \emptyset$ is used then both formulas are completely equal. Furthermore it is a special case of the ESS method, i.e. for all states s_1, s_2 in A_G, if there exists a consistent set of equality assumptions m, such that $F_{\preccurlyeq}(s_1, s_2, m)$ holds, then $F_{\preccurlyeq}(s_1, s_2, m)$ implies $ESS_{\preccurlyeq}(s_1, s_2)$. This follows immediately by definition. Whenever a value can be constructed in s_1, the equality assumptions m can be used to transfer it to s_2. This is always possible, because m is consistent. Section A.1

in the Appendix presents two algorithms that can be used to generate sets of consistent equality assumptions. Every one of these sets can be used as argument to the formula $F_{\preccurlyeq}$ or $F_{\approx}$ to detect (symbolic) state subsumption or equivalence.

This extended SBC method is now able to handle the *fresh symbolic literal problem*. The symbolic state parts $(pc\colon T, a\colon x_1)$ and $(pc\colon T, a\colon x_2)$ of the *tokenring* example are now detected to be equivalent using the equality assumptions $m = \{(x_1, x_2)\}$. Doing so results in the negated formula $\neg F_{\approx}(s_1, s_2, m) = [(x_1 = x_2) \wedge ((T \neq T) \vee (x_1 \neq x_2))]$, which is unsatisfiable. Another example is $\zeta(s_1)=(pc\colon T, a\colon x_1 * x_2 + x_1)$ and $\zeta(s_2)=(pc\colon T, a\colon y_2 * (y_1 + 1))$. Using the equality assumptions $m_1 = \{(x_1, y_1), (x_2, y_2)\}$, the formula $\neg F_{\approx}(s_1, s_2, m_1)$ is satisfiable with the model $\psi = \{x_1 = 0, x_2 = 1, y_1 = 1, y_2 = 0\}$. But $\neg F_{\approx}(s_1, s_2, m_2 = \{(x_1, y_2), (x_2, y_1)\})$ is unsatisfiable. Thus the states are equivalent.

Remarks on consistent equality assumptions Using no equality assumptions at all, keeps the method sound, but fails to detect many equivalent states due to the usage of fresh symbolic literals. On the other hand, assuming too many equalities between symbolic literals can lead to false positives. The reason is they may restrict the possible values that the variables of the states may assume during the solver-based equivalence check, e.g. consider the following example with two states s_1 and s_2 and their linked values v and w and unconstrained path conditions: $v_1 = x_1$, $w_1 = x_2$, $v_2 = x_1$, $w_2 = x_1$. When using the equality assumptions $m = \{(x_1, x_2)\}$ the formula

$$F_{\preccurlyeq}(s_1, s_2, m) = ((x_1 = x_2) \wedge T) \implies (\neg T \vee (x_1 \neq x_1) \vee (x_2 \neq x_1))$$

is unsatisfiable, which incorrectly means that s_1 is covered by s_2. This result is incorrect, because when the states are executed independently, then $v_1 = 0$ and $w_1 = 1$ is a possible valuation in s_1, which has not been considered in the equality check, since the equality assumptions have prevented it. Now if the statement `assert (v == w)` is executed from both states independently, then it will fail in s_1 but pass in s_2. This behaviour clearly contradicts with the definition of $s_1 \preccurlyeq s_2$.

The idea is to only allow equality assumptions, that do not restrict the values that the variables can assume when the states would be executed independently (to show $s_1 \preccurlyeq s_2$ it is sufficient to allow all possible values of s_1). This ensures, that every possible value is considered by the solver, when searching for a counter example that makes the states unequal. For this reason it is required, that the equality assumptions are consistent (unambiguous and type-compatible). Consistent equality assumptions do not restrict the values that variables of a state can assume.

Another example, that is correctly rejected as non-equivalent, is $\zeta(s_1)=(pc\colon T, a\colon x_1 * x_2 + x_1,$ $b\colon x_1)$ and $\zeta(s_2)=(pc\colon T, a\colon y_2 * (y_1 + 1), b\colon y_1)$. It becomes necessary, that x_1 is mapped on y_1[4] for the linked variables b_1 and b_2 to be equal. For a_1 and a_2 to be equal, the equality assumptions $\{(x_1, y_2), (x_2, y_1)\}$ are required. So all together the equality assumptions $m=\{(x_1, y_1), (x_1, y_2),$ $(x_2, y_1)\}$ are required, such that $F_{\approx}(s_1, s_2, m)$ becomes unsatisfiable. But m is inconsistent (since x_1 is mapped on both y_1 and y_2, so it will not be considered. A consistent set of equality assumptions m', that makes $F_{\approx}(s_1, s_2, m')$ unsatisfiable, does not exist for this example. So these states would be correctly classified as non equivalent (e.g. executing `assert (b != 0` `|| a == 0)`; would pass in s_1 but fail in s_2).

---

[4]Actually the mapping direction does not matter, since by definition equality assumptions are symmetric (since equality is symmetric).

```
1    int NUM_ITERATIONS = 3;              34
2    int result = 0;                      35   int fB() {
3    int step = 0;                        36     int b1 = ?(int);
4    event e;                             37     int b2 = ?(int);
5                                         38     int b3 = b2*(b1 + 1);
6    bool is_even(int x) {                39
7      return x % 2 == 0;                 40     assume pred(b2);
8    }                                    41
9                                         42     return b3;
10   bool pred(int x) {                   43   }
11     return (x == 2) || (x == 4);       44
12   }                                    45   thread B {
13                                        46     while (true) {
14   int fA() {                           47       if (is_even(step)) {
15     int a1 = ?(int);                   48         result = result + fB();
16     int a2 = ?(int);                   49         notify WRITE, 0;
17     int a3 = a1*a2 + a1;               50       }
18                                        51       step += 1;
19     assume pred(a1);                   52       wait READ;
20                                        53     }
21     return a3;                         54   }
22   }                                    55
23                                        56   thread check {
24   thread A {                           57     while (true) {
25     while (true) {                     58       wait WRITE;
26       if (is_even(step)) {             59       notify READ, 1;
27         result = result + fA();        60     }
28         notify WRITE, 0;               61   }
29       }                                62
30       step += 1;                       63   main {
31       wait READ;                       64     start (NUM_ITERATIONS - 1);
32     }                                  65     assert is_even(result);
33   }                                    66   }
```

Listing 6.7: *symmetric-accumulator*, example program that generates multiple semantically equivalent but structurally different states.

6.5.1. Example: Symmetric Accumulator

An example program, called *symmetric-accumulator*, that demonstrates the usefulness of a solver-based state comparison with equality assumptions is shown in Listing 6.7. This program is safe. It runs for a specified number of iterations. Thus cycle detection is not necessary to prove it safe, but it does improve the performance considerably by detecting equivalent states due to different scheduling interleavings. The program computes a global result. In each iteration the result is updated by either thread A or B, depending which is executed first, with a locally produced symbolic value. These local values are semantically equivalent but structurally different. In each step fresh symbolic literals are introduced. The thread *check* asserts that the stepwise produced result is valid. Then it starts the next iteration by enabling both threads A and B and waits until the result value is updated (notification from A or B). The interleavings of A and B are not reduced with partial order reduction (even dynamic), because the step variable is always read in both threads and one of them always increments it (writes to it).

A partial order reduced state space of the program is shown in Figure 6.3. The states are numbered in the order a depth first search would visit them[5]. Three state pairs are equivalent: $s_3 - s_{16}$, $s_6 - s_{14}$ and $s_{12} - s_9$. Their symbolic state parts are shown in Table 6.1. The value of the step variable is also displayed for convenience. The terms $x_1..x_6$ and $y_1..y_6$ denote the symbolic literals, which are introduced in the lines $\{15, 16, 36, 37\}$ during the execution of the program. The equality assumptions m that lead to the detection of the equivalent state pairs are:

---

[5]The initial transition is arbitrary chosen, though a different choice would lead to a similar state exploration diagram, but the subsequent transitions follow from the initial decision.

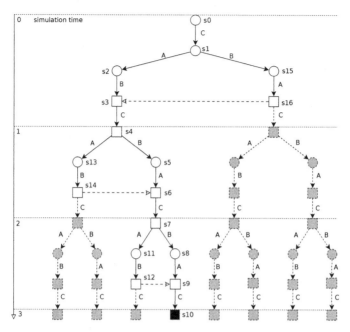

Figure 6.3.: Complete partial order reduced state space for the program in Listing 6.7 for three time steps (0,1,2). Due to POR it is only necessary to consider interleavings of A and B, since C is independent to both of them (in the initial state s_0 and later they are no longer co-enabled). Circles denote normal states, squares represent states without enabled threads so they perform a delta or timed notification. All squares at the bottom represent terminal states. The dashed connections between $s_3 - s_{16}$, $s_6 - s_{14}$ and $s_{12} - s_9$ indicate that these state pairs are equivalent. The direction of the arrow points to the already visited state. The dotted gray shaded states (connected with dotted lines) are not explored since they are successors of equivalent states. This state space could be generated, if a depth first search executes the path from s_0 to s_{10} first. This diagram assumes, that simulation starts with thread C, starting with either A or B would result in a similar state space exploration diagram.

Table 6.1.: Relevant state parts for the detection of the equivalent states in program Listing 6.7 as shown in Figure 6.3

state	data	value
s_3	step	2
	pc	$(x_1 = 2 \vee x_1 = 4)$
	result	$0 + (x_1 * x_2 + x_1)$
s_{16}	step	2
	pc	$(y_2 = 2 \vee y_2 = 4)$
	result	$0 + (y_2 * (y_1 + 1))$
s_6	step	4
	pc	$(x_1 = 2 \vee x_1 = 4) \wedge (y_4 = 2 \vee y_4 = 4)$
	result	$0 + (x_1 * x_2 + x_1) + (y_4 * (y_3 + 1))$
s_{14}	step	4
	pc	$(x_1 = 2 \vee x_1 = 4) \wedge (x_3 = 2 \vee x_3 = 4)$
	result	$0 + (x_1 * x_2 + x_1) + (x_3 * x_4 + x_3)$
s_9	step	6
	pc	$(x_1 = 2 \vee x_1 = 4) \wedge (y_4 = 2 \vee y_4 = 4) \wedge (y_6 = 2 \vee y_6 = 4)$
	result	$0 + (x_1 * x_2 + x_1) + (y_4 * (y_3 + 1)) + (y_6 * (y_5 + 1))$
s_{12}	step	6
	pc	$(x_1 = 2 \vee x_1 = 4) \wedge (y_4 = 2 \vee y_4 = 4) \wedge (x_5 = 2 \vee x_5 = 4)$
	result	$0 + (x_1 * x_2 + x_1) + (y_4 * (y_3 + 1)) + (x_5 * x_6 + x_5)$

$$s_3 \simeq s_{16} \text{ with } m_1 = \{(x_1, y_2), (x_2, y_1)\}$$
$$s_6 \simeq s_{14} \text{ with } m_2 = \{(x_3, y_4), (x_4, y_3)\}$$
$$s_9 \simeq s_{12} \text{ with } m_3 = \{(x_5, y_6), (x_6, y_5)\}$$

6.5.2. Generating Consistent Equality Assumptions

As already mentioned, Section A.1 in the Appendix presents two algorithms in detail, that can be used to calculate sets of consistent equality assumptions. The first algorithm is conceptually simpler and will generate all possible consistent sets, whereas the second algorithm will only generate a subset of them by filtering out sets of equality assumptions, that most likely will not lead to the detection of state subsumption. The first algorithm serves to illustrate the basic concepts, but it is not scalable, thus it will not be further considered. The second algorithm is the one that has been implemented. This section summarizes the functionality of the second algorithm and illustrates it by deriving the equality assumptions $m_2 = \{(x_3, y_4), (x_4, y_3)\}$ for the *symmetric accumulator* example program of the previous section. These equality assumptions have led to the detection, that the states s_6 and s_{14} are equivalent. The equality assumptions for the other equivalent state pairs can be derived analogously.

Algorithm Summary

The algorithm consists of three subsequent phases: a preprocessing-, a construction- and a generation- phase. The first two phases create an intermediate representation, namely a set of (equality) constraints. The generation phase then generates sets of consistent equality assumptions from it. Every of these sets can be used as argument to the formula $F_{\preccurlyeq}$ or $F_{\sim}$ to detect (symbolic) state subsumption or equivalence. A constraint is a pair of symbolic literal sets (A,B). It denotes that any literal $a \in A$ can be mapped upon any literal $b \in B$ (or the other way round since equality is symmetric). Constraints are reduced into smaller ones by splitting them into multiple disjunct parts. This process is denoted as *separation*. For example (A,B) can be separated into (A_1,B_1) and (A_2,B_2) where $A_1 \cap A_2 = \emptyset$ and $B_1 \cap B_2 = \emptyset$. Smaller constraints result in the generation of a smaller number of equality assumption sets, which in turn leads to a smaller number of queries to the SMT solver, thus improving the state matching efficiency. However, too aggressive separation can reduce the state matching precision.

The preprocessing phase constructs the initial constraints and separates each of them independently of the other ones yielding a set of preprocessed constraints. Basically it works by considering every corresponding value pair (v_1, v_2) between the symbolic state s_1 and s_2 one after another in isolation. Each of them is transformed into an initial constraint $c = (A,B)$, where A and B are assigned all reachable symbolic literals from v_1 and v_2 respectively. These initial constraints will be separated, if $A \cap B \neq \emptyset$ or some symbolic literals in A are type incompatible with some symbolic literals in B. The construction phase than further separates the set of preprocessed constraints by analyzing them all together yielding a simplified set of constraints. Basically two constraints (A_1,A_2), (B_1,B_2) will be further separated into smaller constraints, if any one of $A_1 \cap A_2$, $A_1 \cap B_2$, $B_1 \cap A_2$, or $B_1 \cap B_2$ is not disjunct. The generation phase than yields sets of consistent equality assumptions, by mapping every $a \in A$ to a $b \in B$ for every simplified constraint (A,B). More details, including complete algorithm listings, are presented in the Appendix in Section A.1.

Example

As an example, the equality assumptions $m_2 = \{(x_3,y_4), (x_4,y_3)\}$, that have led to the detection of s_6 and s_{14} to be equivalent for the *symmetric accumulator* example program in the previous section, will be derived in the following using the (second) algorithm, which has been summarized above.

First a set of preprocessed constraints is constructed. Both symbolic state parts have two entries, that can be grouped to pairs: the path conditions and the values of the *result* variables. First the pair $(pc(s_6), pc(s_{14}))$ is considered. Its initial constraint (A,B) is computed as $A = \Gamma(pc(s_6)) = \{x_1,y_4\}$ and $B = \Gamma(pc(s_{14})) = \{x_1,x_3\}$. Since A and B neither contain any type incompatible nor equal literals, the constraint is kept without further separation. Next the pair $(vars(s_6)[result], vars(s_{14})[result])$ is considered. In this case $A = \{x_1,x_2,y_3,y_4\}$ and $B = \{x_1,x_2,x_3,x_4\}$ which is separated into the three constraints $(\{x_1\}, \{x_1\})$, $(\{x_2\}, \{x_2\})$ and $(\{y_3,y_4\}, \{x_3,x_4\})$, since $A \cap B = \{x_1,x_2\}$. The set of preprocessed constraints for both pairs is then $\{c_1 = (\{x_1,y_4\}, \{x_1,x_3\}), c_2 = (\{x_1\}, \{x_1\}), c_3 = (\{x_2\}, \{x_2\}), c_4 = (\{y_3,y_4\}, \{x_3,x_4\})\}$.

These preprocessed constraints are then considered one after another, to construct a set of simplified constraints. W.l.o.g. the constraints are considered in the order $c_1..c_4$ during the construction phase. Initially the result set is empty. First c_1 is simply added to the result set. Next c_2 is analyzed. It will result in the separation of c_1 into $(\{x_1\}, \{x_1\})$ and $(\{y_4\}, \{x_3\})$. Then c_3 is analyzed and simply added to the result set, since it has no common symbolic literals with any so far collected constraint. Lastly the constraint c_4 is analyzed and separated into $(\{y_4\}, \{x_3\})$ and $(\{y_3\}, \{x_4\})$, due to the already collected constraint $(\{y_4\}, \{x_3\})$. The resulting set of simplified constraints is $\{(\{y_4\}, \{x_3\}), (\{x_1\}, \{x_1\}), (\{x_2\}, \{x_2\}), (\{y_3\}, \{x_4\})\}$.

Remark. The result would be the same, even if the order in which the preprocessed constraints are considered would be changed, e.g. if c_4 is added before c_2, then c_1 and c_4 would be separated into $(\{x_1\}, \{x_1\})$, $(\{y_4\}, \{x_3\})$, $(\{y_3\}, \{x_4\})$. Then c_2 would be discarded, since $(\{x_1\}, \{x_1\})$ is already available. Lastly c_3 would normally be collected without further separating any collected constraints, thus yielding the same result.

The generation phase then yields only a single set of equality assumptions, namely $m = \{(y_4, x_3), (y_3, x_4)\}$. The implicit equality assumptions (x_1, x_1) and (x_2, x_2) are not part of m, since they are equal anyway. Only explicit equality assumptions are kept[6]. Using m, the formula $F_{\simeq}(s_5, s_{13}, m)$ is unsatisfiable, thus s_6 and s_{14} are equivalent.

6.5.3. Breaking Implicit Equality Assumptions

This section presents an optimization that can increase the detection rate of the extended solver-based algorithm. Its based on the observation that often symbolic values are shared between multiple execution paths, which are represented by execution states, as an optimization. Symbolic literals that are shared cannot be mapped upon other symbolic literals, since they are already implicitly equal between both execution states, thus every consistent set of equality assumptions will leave them unmapped. Replacing all symbolic literals that are shared between two states can improve the detection rate of the extended solver-based algorithm. The detection rate is never reduced, since the implicit equality assumptions can still be assumed, but they are no longer mandatory.

Consider two states s_1 and s_2 with two symbolic variables v and w defined as $v_1 = x_1 + x_2$, $w_1 = x_1$, $v_2 = x_1 + x_2$, $w_2 = x_2$. Both path conditions are unconstrained. These two symbolic state parts are equivalent. But they won't be detected as such with the current methods. The problem is, that x_1 and x_2 are symbolic literals that are shared on both states (e.g. because they originated from a common ancestor state). Thus they are implicitly equal (the solver must assign them the same value, since they are the same entity). So they cannot be mapped on another symbolic literal. Doing so would result in an inconsistent set of equality assumptions. But a consistent mapping is required, to guarantee that the state matching is sound (else it may report false positives[7]). Now if the symbolic literals x_1 and x_2 are replaced with fresh once in e.g. s_1, resulting in $v_1 = x_3 + x_4$ and $w_1 = x_3$, the states s_1 and s_2 will be correctly matched with the equality assumptions $\{(x_2, x_3), (x_1, x_4)\}$.

The disadvantage of this method is its additional runtime overhead. First it must detect common symbolic literals and replace them in one state with fresh ones (or just replace all symbolic literals of one state to avoid detecting common literals). Second and probably more importantly, it normally does increase the number of possible symbolic literal mappings (but which in turn can result in a higher detection rate).

Remark. Introducing fresh symbolic literals and thus circumventing inconsistent mappings will not lead to false positives, so the method is still sound. It will only work out for states that in fact are equivalent. Those which are not, will also be rejected as such even when fresh symbolic literals are introduced.

Consider for example two states with symbolic values: $v_1 = 0 * x_1 + x_2$, $w_1 = x_2$, $v_2 = x_1$, $w_2 = x_2$. These symbolic state parts are clearly not equivalent, since executing the statement *assert (v == w)* from each state independently will be pass for the first state (since v_1 always equals w_1) but fail for the second state (since v_2 may be unequal to w_2). Introducing fresh symbolic literals in s_2 will result in $v_2 = x_3$ and $w_2 = x_4$. So x_4 is mapped on x_2 hence x_3 cannot be mapped on x_2 too, as this would result in an inconsistent mapping. Thus x_3 is mapped on x_1.

---

[6]An equality assumption (a, b) is denoted *implicit* if $a = b$ else it is called *explicit*.

[7]It may detect to states as equivalent, even though they are not.

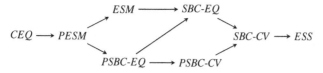

Figure 6.4.: The precision of different state matching methods depicted as partial order (CEQ is the weakest and ESS the strongest method)

The formula

$$(x_1 = x_3 \wedge x_2 = x_4 \wedge pc(s_1)) \wedge (((0 * x_1 + x_2) \neq x_3) \vee (x_2 \neq x_4) \vee \neg pc(s_2))$$

that is generated to check whether to symbolic state parts must be equal (see Definition 27) is satisfiable (and other symbolic literal mappings are not possible). So the states are still correctly rejected as non-equivalent after introducing fresh symbolic literals.

6.6. Classification of State Subsumption Algorithms

Two different heuristic algorithms have been introduced to detect comparable symbolic state parts. The Explicit Structural Matching (ESM) in Section 6.1 and the Solver-Based Comparison (SBC) in Section 6.2. Both versions are available in a plain version, denoted as PESM and PSBC, and an extended version. The normal abbreviations ESM and SBC will refer to the extended version. The extended versions can handle the *fresh symbolic literals* problem, that e.g. appears in the *token-ring* program, as described in Section 6.3. A special variant of the PESM method is the CEQ method. It is equal to PESM except no expression simplifications are employed at runtime. The SBC can be used to detect equivalent states (P)SBC-EQ or covered/subsumed states (P)SBC-CV. The SBC-CV is the strongest of the heuristic methods, but it is still weaker than the solver-based Exact Symbolic Subsumption (ESS) method, that has been presented in Section 5.4.2 of Chapter 5. All other algorithms can be reduced upon the ESS algorithm, but the other way round is not possible. The precision of these methods forms a partial order, which is graphically shown in Figure 6.4. An edge from A to B means that A is strictly weaker algorithm than B, denoted as $A \preceq B$. The following overview provides more informations for every pair of algorithms shown Figure 6.4.

- CEQ $\preceq$ PESM : By definition, PESM is stronger than CEQ, since CEQ is not using expression simplifications. Thus it is unable to handle the *condition-builder* example (Listing 6.3), since it is unable to detect the cycle of states without expression simplification rules. The PESM method on the other hand can verify it.

- PESM $\preceq$ ESM : By definition, PESM can only match equal symbolic literals, whereas ESM can match type compatible symbolic literals. PESM fails on the *tokenring* example (Listing 6.6), but ESM can handle it.

- PSBC-EQ/CV $\preceq$ SBC-EQ/CV : By definition, the PSBC versions do not use additional equality assumptions. None of PSBC-EQ/CV can handle the *tokenring* example, but both SBC-EQ and SBC-CV support it.

- PSBC-EQ $\preceq$ PSBC-CV : By definition, PSBC-EQ requires that both path conditions are equal, while for PSBC-CV it is sufficient if one implies the other. PSBC-CV can detect many state subsumptions when simulating the *symbolic counter* example (Listing 6.5) whereas PSBC-EQ cannot, e.g. $(pc : x_1 < 3, c : x_1 + 1) \preceq (pc : T, c : x_1 + 1)$.

- SBC-EQ $\preceq$ SBC-CV : Analogous to the above case.

- ESM $\preceq$ SBC-EQ : The symbolic literals matched by the ESM algorithm always form a subset m of all possible consistent equality assumptions. The SBC-EQ algorithm is compatible with all consistent equality assumptions. Thus if the state parts do structurally match with m, the solver will detect it too using the equality assumptions m. On the other hand, ESM misses many equivalent states in the *symmetric accumulator* example (Listing 6.7) and fails to prove the time unbounded version of the *rotating buffer* example (Listing 6.4) to be safe. Though the validity of these results depends on the simplification/normalization engine used. A simple example that will always fail with ESM but pass with SBC-EQ is $(pc : T, a : 5) \simeq (pc : (x_1 = 5), a : x_1)$.

- PESM $\preceq$ PSBC-EQ : Analogous to the above case.

- SBC-CV $\preceq$ ESS : Whenever a subsumption of two states is detected by the SBC-CV method, there exists a set of consistent equality assumptions m such that all variable values are equal and the path condition of the subsumed state implies the path condition of the other state. Thus the ESS method will detect the subsumption too. On the other hand, SBC-CV will not detect the subsumption $(pc : T, a : 2 * x_1) \preceq (pc : T, a : x_1)$, whereas the ESS method will detect it.

The actual performance of the methods does not have to correspond to their precision, as shown in Figure 6.4, though. So e.g. the runtime of the SBC methods can be slower than the runtime of the ESM methods, due to the more complex solver queries. The extended versions, compared to their base versions (e.g. ESM compared to PESM), offer a greater precision with negligible runtime overhead for those cases, which both methods can handle. Thus the extended versions seem to be always superior to the base methods (the ESM is even better for all cases, but the SBC can spend longer time building equality assumptions - but for example where no fresh symbolic literals are used, there is no need for equality assumptions thus SBC would perform the same way as PSBC). Detecting covered states can further improve the performance of a stateful state exploration, as the (already presented) *symbolic counter* example in Listing 6.5 suggests. Though proving an implication might be a harder problem for the solver, than proving equality.

6.7. Further Optimization

This section sketches some promising optimizations, that have not yet been implemented.

6.7.1. Garbage Collecting Path Condition Constraints

The state comparison methods described so far require that the path conditions of the compared states s_1, s_2 are compatible (either equal or one implies the other). During symbolic execution it regularly happens that both branch successors are feasible. In such a case the execution of a path s will fork into two paths s_T and s_F. Their path conditions will be extended accordingly, with the branch condition c (s_T) or its negation (s_F). Thus $pc(s_T) \wedge pc(s_F)$ is unsatisfiable. So these forked paths (and any of their descendants, as they inherit the path condition) will never

```
1   int f() {                           9
2       int z = ?(int);                10   thread A() {
3       if (z < 5) {                   11       int a = f();
4           return 1;                  12       ...
5       } else {                       13       a = abs(a);
6           return -1;                 14       wait_time 0;
7       }                              15       ...
8   }                                  16   }
```

Listing 6.8: Code snippet that illustrates the need for path condition simplification

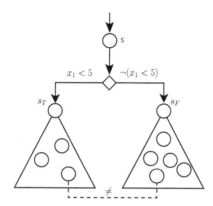

Figure 6.5.: Neither the forked paths s_T and s_F, nor any of their corresponding descendants (represented by the cones) will be detected as equivalent, without simplifying their path condition.

be detected as equivalent. The principle is illustrated in Figure 6.5. Even though some of these descendant states are equivalent, they will not be detected as such, since the path conditions are not simplified. They keep all their constraints even if the symbolic literals involved can no longer influence the program execution. Listing 6.8 shows a code snippet that can be used to construct such a situation.

When thread A is executed, it will call function f, which will create a fresh symbolic literal x_1 and assign it to the local variable z. Since z is unconstrained, both paths of the following conditional branch are feasible. Thus the execution of thread A will fork and both paths will be continued independently, with accordingly extended path conditions, s_T ($x_1 < 5$) and s_F ($\neg(x_1 < 5)$). Once the function f returns, the local variable z will be destroyed and its contained symbolic literal x_1 becomes unreachable.

The result of f will be stored in the local variable a (either 1 or -1) of thread A. Then later a will be overwritten with its absolute value, thus the value of a becomes equal (only 1) on both forked paths (and their possible descendants that might have been spawned between Line 11 and Line 13). Now assuming that the states of s_T and s_F have only been modified uniformly between Line 11 and Line 13, then they are both completely equal except for their path conditions, which are still incompatible: $x_1 < 5 \in pc(s_T)$ and $\neg(x_1 < 5) \in pc(s_F)$.

But x_1 is unreachable in both states. And in neither path condition it is part of an expression which contains reachable symbolic literals. This means it cannot longer be referenced from the program execution and cannot influence the possible values of other reachable symbolic literals in the path condition. Thus the whole terms $x_1 < 5$ and $\neg(x_1 < 5)$ can be safely removed (or replaced with the concrete value *true*) from the corresponding path condition. Doing so would allow the scheduler to detect that both paths are equivalent at the context switch (Line 14). Based on these observations the following definition seems to be valid.

Definition 28

> A *n-ary logic expression (condition) e in the path condition can be replaced with the concrete literal* true, *if neither argument of e contains a reachable symbolic literal. A symbolic literal is reachable, if there exists any reachable reference to it, starting from any local or global variable value.*

The basic idea is summarized in Definition 28. To simplify the path condition it is necessary to detect unreachable values. This problem is part of a typical garbage collection (G.C.) method. Many different G.C. algorithms have been proposed in the literature [TC11; Aho+06]. Two well-known methods are sketched in the following.

A conceptually simple method would be to periodically (and preferably incrementally) collect a set of all reachable symbolic literals starting from each local and global variable slot. All values not in this set are considered unreachable. This garbage collection. method is commonly known as *mark and sweep*.

Another possibility is to keep a reference counter for each value. Whenever a value is overwritten or destroyed, its counter will be decreased. When it hits zero, then all reference counters of all reachable values will also be decreased. A symbolic literal with a zero count is considered unreachable. Reference count methods need a way to handle unreachable reference cycles, which prevents any node of the cycle to reach a zero count.

6.7.2. Combining the Explicit and Solver-Based Method

The possible number of generated symbolic literal mappings can be exponential to the maximum number of matched literals. And for every mapping an SMT formula will be generated and checked for satisfiability, which itself is a hard task. This is repeated until an unsatisfiable formula is detected (which means both states are equivalent or one is covered by the other), the number of mappings is exhausted or some heuristic requests abortion, because too many mappings have already been checked. Thus it is crucial to keep the number of considered mappings low and select those mappings first, which seem most probable to produce an unsatisfiable formula.

This sections shows another method that aims to improve the former requirement. The idea is to combine the explicit with the solver-based method. Given two states, s_1 and s_2 that shall be compared, every pair (v_1, v_2) of linked symbolic values between s_1 and s_2 is matched using the ESM (Explicit Structural Matching) method. Each of these matching applications can either succeed or fail for one of two reasons: either a structural mismatch is detected between v_1 and v_2 or an inconsistent mapping is generated. This concept is illustrated in Figure 6.6.

If the ESM finishes for all values without generating an inconsistent mapping, then all of its collected (one-to-one) constraints are added to the initial complete constraint list generated by the solver-based method. The SBC (Solver-Based Comparison) method will proceed normally as already described in Section A.1, e.g. by applying type- or location-based grouping to split existing constraints into smaller ones. Else the complete state comparison aborts and both states will be considered non equivalent. Aborting in this case (an inconsistent mapping has been generated) is just another heuristic employed here. Actually the solver-based method as described in Section A.1 can handle such constraints correctly (by splitting them and thus assuming no equality). But normally such states won't be equal anyway, so it makes sense to

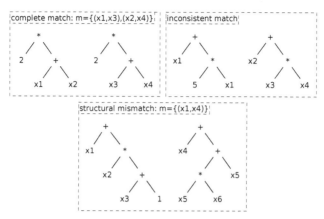

Figure 6.6.: The upper left expression trees for the values $v_1 = 2 * (x_1 + x_2)$ and $v_2 = 2 * (x_3 + x_4)$ can be completely matched. Doing so will generate the two constraints $(\{x_1\}, \{x_3\})$ and $(\{x_2\}, \{x_4\})$. The upper right trees for the values $v_1 = x_1 + (5 * x_1)$ and $v_2 = x_2 + (x_3 * x_4)$ will result in an inconsistent mapping, since the symbolic literal x_1 is matched with both x_2 and x_4. The expression trees in the lower diagram for the values $v_1 = x_1 + (x_2 * (x_3 + 1))$ and $v_2 = x_4 + ((x_5 * x_6) + x_5)$ can be partially matched. Only one constraint $(\{x_1\}, \{x_4\})$ will be generated.

filter them out early.

Remark. The ESM can only add constraints which are subset of the complete constraints generated by the SBC. Thus it can never increase the number of possible symbolic literal mappings, only decrease them.

Example 3. Assuming two states s_1 and s_2 shall be matched, and both of them have unconstrained path conditions and two linked symbolic values v and w, that correspond to the upper left and lower middle examples of Figure 6.6. So $v_1 = 2 * (x_1 + x_2)$, $v_2 = 2 * (x_3 + x_4)$, $w_1 = y_1 + (y_2 * (y_3 + 1))$, $w_2 = y_4 + (y_5 * y_6 + y_5)$. Further assuming that only type based grouping is employed and all symbolic literals of the pairwise linked values have the same type, then the solver-based method alone would generate the following constraints:

$(\{x_1, x_2\}) - (\{x_3, x_4\})$
$(\{y_1, y_2, y_3\}) - (\{y_4, y_5, y_6\})$

Thus it would be possible to generate 12 (complete) symbolic literal mappings using the (second) algorithm presented in Section A.1.3. But with the additional constraints $(\{x_1\}, \{x_3\})$, $(\{x_2\}, \{x_4\})$, $(\{y_1\}, \{y_4\})$ collected by the ESM preprocessing, the resulting constraints would be:

$(\{x_1\}) - (\{x_3\})$
$(\{x_2\}) - (\{x_4\})$
$(\{y_1\}) - (\{y_4\})$
$(\{y_2, y_3\}) - (\{y_5, y_6\})$

Doing so would reduce the number of (complete) mappings to 2, namely $m_1 = \{(x_1, x_3), (x_2, x_4), (y_1, y_4), (y_2, y_5), (y_3, y_6)\}$ and $m_2 = \{(x_1, x_3), (x_2, x_4), (y_1, y_4), (y_2, y_6), (y_3, y_5)\}$.

7. Experiments

In this thesis a complete symbolic simulation has been proposed to verify SystemC programs with arbitrary finite state space. Two complementary optimization techniques, namely POR (Chapter 3-4) and SSR (Chapter 5-6), have been integrated to alleviate the state explosion problem and thus allow for the verification of non-trivial programs. Both, POR and SSR, are available in different variants. This section compares the different configurations with each other and with state of the art tools in the verification of SystemC programs.

The experiments are performed on a Linux PC with a 1.4 GHz Intel Dual Core and 6GB ram. The time and memory limits are set to 500 seconds and 2GB respectively. The abbreviations T.O. and M.O. denote that the time and memory limit has been exceeded, respectively. The abbreviation N.S. denotes that a program is not supported by a specific configuration. This happens when a stateless algorithm is applied on a cyclic state space, where not every execution path eventually runs into an error state, or a SSR method is used that is unable to detect a relevant cycle in the program. The abbreviation N.A. denotes that some information is not available for a specific configuration, e.g. the number of state matches in a stateless search. The runtimes are specified in seconds.

The rest of this chapter is structured as follows: First the different configurations of POR and SSR are presented. Then the benchmarks used in the subsequent experiments will be shown. Next the effect of symbolic state part hashing in the ESM method is evaluated. Thereafter the different supported solvers are compared, followed by the different POR variants and state matching methods. Based on the observed results some configurations will be selected for the main benchmark, that compares these methods against the *Kratos* model checker [CNR13]. As already mentioned, *Kratos* represents one of the current state of the art approaches in SystemC verification.

7.1. Configuration Overview

An overview of the different configurations is shown in Figure 7.1. Any configuration in a leaf can be selected for POR and SSR.

Two different stateful explorations have been implemented based on static and dynamic POR, respectively. The former is based on the AVPE algorithm, as described in Section 3.2.3. The

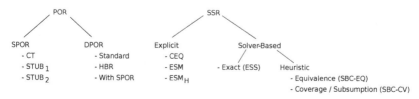

Figure 7.1.: Component overview

latter is based on the SDPOR algorithm, as described in Section 4.4. Thus in the following all stateful explorations with static POR use the AVPE algorithm and those with dynamic POR use the SDPOR algorithm. The SDPOR algorithm can optionally use a *happens before* relation (HBR) to infer smaller persistent sets and fall back to static persistent sets whenever the dynamic inference of a non trivial persistent set fails.

Static persistent sets can be computed using either the *conflicting transitions* (CT) or *stubborn set* algorithm. The latter is available in two different configurations denoted as $STUB_1$ (or simply $STUB$) and $STUB_2$. They differ in the way how *necessary enabling transitions* are computed whenever a dependency to a disabled transition is detected during the computation of the persistent set. All of these three algorithms can compute a minimal persistent set locally for each state. This extension will simply compute all possible persistent sets for a given state (one for each enabled transition) and then select the smallest one. The suffix *-MIN* will be used to denote that a minimal persistent set is computed. Thus the configuration $STUB\text{-}MIN$ means that the smallest persistent set shall be computed using the *stubborn set* algorithm.

Different state matching predicates have been implemented as described in Chapter 6 for the State Subsumption Reduction (SSR). Basically they can be divided in explicit and solver-based methods. The former are more efficient (polynomial worst case complexity) but less precise than the latter. A classification of the methods is available in Section 6.6.

The explicit methods are the Complete Equality Matching (CEQ) and the Explicit Structural Matching (ESM). The former requires that two states are completely equal, whereas the latter involves expression simplification rules, and can handle the *fresh symbolic literals problem*, as introduced in Section 6.3, compared to CEQ. The ESM can be seen as a generalization of the CEQ method. The ESM method can optionally include symbolic state parts into the computation of a state's hash value, as described in Section 6.1.3. This variant of ESM will be denoted as ESM_H. It can result in a smaller number of unnecessary state comparisons.

The solver-based (SBC) methods are the heuristic (symbolic) equivalence (SBC-EQ), heuristic (symbolic) subsumption/coverage (SBC-CV) and exact (symbolic) subsumption/coverage (ESS). All of them query an (SMT) solver to decide whether two states match. The simulator supports different solvers through *metaSMT* [Hae+11] and direct access of the Z3 solver. The ESS method is always used in combination with the direct Z3 solver, since it is currently the only supported solver that provides support for quantifiers, as required by the ESS method. The other methods can use any of the available solvers.

Remark. The heuristic SBC methods require an algorithm that computes consistent equality assumptions. Two different algorithms have been presented in Section A.1. The first one has already been identified as not scalable, thus it will not be further considered. The second algorithm itself can be further configured. Symbolic literals can either be matched by type or by their program location, when generating constraint sets from corresponding (symbolic) variable pairs. A constraint set consists of two sets of symbolic literals (A, B) and denotes that any symbolic literal in A can be mapped upon any other symbolic literal in B. Assuming w.l.o.g. $|A| < |B|$, all symbolic literals in A will be mapped on exactly one (unique) symbolic literal in B, resulting in a set of consistent equality assumptions. The algorithm can either accept all constraints or only regular ones, as described in Section 11. A regular constraint is one where $|A| = |B|$. Accepting all constraints in combination with type matching is the most precise heuristic configuration, whereas matching by program location and accepting only regular constraints should result in fewer unnecessary sets of equality assumptions. Experimental evaluation of these additional heuristics is postponed for future work. In the following the default configuration, which is matching by type and accepting all constraints, is used.

Table 7.1.: List of (unparameterized) benchmarks. Some benchmarks will be abbreviated with the optional short name, due to page size constraints.

Benchmark	Short name	Version	Source
bist-cell		S	[CNR13]
buffer		S	[Le+13]
condition-builder	cb	S	Listing 6.3 in Section 6.1.2
counter		S	this section
kundu		S	[KGG08]
kundu-bug-1		U	[CNR13]
kundu-bug-2		U	[CNR13]
mem-slave-tlm	mst	S	[CNR13]
mem-slave-tlm.bug	mst.bug	U	[CNR13]
mem-slave-tlm.bug2	mst.bug2	U	[CNR13]
pc-sfifo-sym-1		S	[CNR13]
pc-sfifo-sym-2		S	[CNR13]
pressure-safe	pres-safe	S	[BK10]
rbuf		S	Listing 6.4 in Section 6.2.1
rbuf2		S	this section
rbuf2-bug1		U	this section
rbuf2-bug2		U	this section
simple-fifo	sfifo	S	[GLD10; Le+13]
simple-pipeline		S	[CNR13]
symbolic-counter	sym-cnt	S	Listing 6.5 in Section 6.2.2
symmetric-accumulator	sym-acc	S	Listing 6.7 in Section 6.5.1
term-counter		S	this section
token-ring		S	[CNR13]
token-ring-bug		U	[CNR13]
token-ring-bug2		U	[CNR13]
toy-sim		S	[CNR13]
transmitter		U	[CNR13]
transmitter-safe		S	safe variant of *transmitter*

Remark. All heuristic methods are available in a plain and extended version as described in Section 6.6. The difference is, that the extended versions can handle the *fresh symbolic literal problem*, as described in Section 6.3. When applied to a benchmark, where the problem does not occur, both methods perform similarly. But the extended versions can also handle programs where fresh symbolic literals are introduced. Thus only the extended versions will be further considered.

7.2. Benchmark Overview

The benchmarks, that will be used for the experiments, are a combination of those already presented in Chapter 6 and from the literature on SystemC verification [KGG08; BK10; GLD10; Cim+11; CNR13; Le+13]. Some benchmarks from the literature have been slightly adapted and a few more benchmarks will be introduced in this section. A complete list of (unparameterized) benchmarks is shown in Table 7.1. The versions denotes whether the benchmark is safe (S) or contains bugs (U). The source shows where the benchmarks have been retrieved. This may not always be the original source. Adapted benchmarks are still attributed to the (original) source. They will be briefly described in the following. Some of the benchmarks are available in different variants and unsafe versions. The *token-ring* program has also been described in this thesis in the context of the *fresh symbolic literal problem* in Section 6.3.

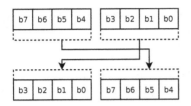 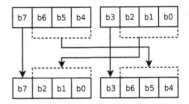

Figure 7.2.: The rotating (bit-) buffer benchmark is available in two configurations *rbuf* and *rbuf2*. In every time step a new value is computed based on the current value of the buffer. The left figure shows the principle operation of the *rbuf* benchmark and the figure on the right for the *rbuf2* benchmark.

Adapted Benchmarks

pressure-safe The program *pressure* appeared in [BK10]. Basically a thread increments a counter while another thread guards the counter from exceeding a maximum value. However, there exists some thread interleavings such that the condition is violated. The *pressure-safe* program limits the simulation time to ensure that the violation does not occur. [BK10] has already observed that both threads are effectively non-interfering most of the time.f However this effective non-interference will normally not be detected, neither by SPOR nor DPOR, because a read-write dependency exists[1]. A stateful search however can avoid re-exploration of the equivalent states.

buffer The *buffer* program first fills an array using multiple threads and then checks the result. All threads write to a different memory locations, thus they are non-interfering. It is based on the *buffer-ws-pX* benchmark from [Le+13] and has been modified to run without a simulation time limit. Consequently, it has a cyclic state space and cannot be verified with a stateless method.

simple-fifo The *simple-fifo* benchmark consists of a configurable number of consumer and producers, that alternately write-to and read-from a shared fifo, which is implemented as an array.

other benchmarks The *toy-sym*, *pc-sfifo-sym-1* and *pc-sfifo-sym-2* benchmark respectively are originally based on the *toy*, *pc-sfifo-1* and *pc-sfifo-2* benchmarks respectively. The concrete counter values have been replaced with symbolic ones, to utilize symbolic state subsumption matching. The *simple-pipeline* benchmark is based on the *pipeline* benchmark. The latter contains unnecessary statements (e.g. notification of events that are never used in the program), which have been removed. The *transmitter-safe* benchmark is a safe version of the *transmitter* benchmark.

New Benchmarks

rbuf variants The rotating bit-buffer program is available in two different configuration. The first *rbuf*, which has already been described in Section 6.2.1, uses either two or four threads to rotate a 32 bit unsigned value c in each step into a new value n. Thus $n[31..16] = c[15..0]$ and $n[15..0] = c[31..16]$ when two threads are used and $n[31..24] = c[23..16]$, $n[23..16] = c[15..8]$, $n[15..8] = c[7..0]$, $n[7..0] = c[31..24]$ for the benchmarks with four threads. The second *rbuf2*

---

[1]The *guard* thread will read the counter in each step, even though it will not modify it most of the time.

```
 1  int STEPS = 50;              16                             31     wait_time 1;
 2  event eA;                    17  thread B {                 32  }
 3  event eB;                    18    while (b < STEPS) {      33  }
 4                               19      b += 1;                34
 5  int a = ?(int);              20      changes += 1;          35  main {
 6  int b = ?(int);              21      wait eB;               36    assume (a == 0 || a ==
 7  int changes = 0;             22    }                                  1);
 8                               23  }                          37    assume (b == 0 || b ==
 9  thread A {                   24                                         1);
10    while (a < STEPS) {        25  thread clk {               38    start;
11      a += 1;                  26    wait_time 1;             39    assert (a == STEPS &&
12      changes += 1;            27    while (changes > 0) {             b == STEPS);
13      wait eA;                 28      changes = 0;           40  }
14    }                          29      notify eA;
15  }                            30      notify eB;
```

Listing 7.1: *term-counter*, example program that generates many states with equal concrete state parts but different symbolic state parts

manages a signed 32 bit value. It rotates all but the first bit of each section. Thus $n[30..16] = c[14..0]$, $n[14..0] = c[30..16]$ and $n[31] = c[31]$, $n[15] = c[15]$ for two threads. The version with four threads is defined analogously. The principle is shown in Figure 7.2. Both of them check that the number of high bits is equal before and after each rotation step. The program is available in with and without a simulation time bound, e.g. the *rbuf-4.5* is the first version with four threads simulated five time steps (a single rotation step is performed in each time step). And e.g. the program *rbuf2-2* is the second version with two threads and no simulation time limit. The *rbuf2* program is also available in two unsafe variants, denoted *rbuf2-bug1.2* and *rbuf2-bug2.2*, which rotate the bits incorrectly in each step.

term-counter The *term-counter* program is shown in Listing 7.1. It is a safe program with acyclic state space. It contains two (independent) threads that count up to a specified limit. A clock thread re-enables them in each time step. The number of threads and the count limit can be parameterized to create different benchmarks. Some results obtained by running this program with 1..4 threads and with 50, 100, 200 and 400 steps, are shown in Table 7.2.

counter The *counter* program is similar to the *term-counter* program. It also has one to four independent threads that count up to a specified limit. But the clock thread, that notifies the counter threads in each step, will not terminate once all threads are done but keeps sending notifications. Thus cycle detection, e.g. by using a stateful search, is required to verify it.

7.3. Comparing Hashing Methods for Symbolic State Parts

A stateful search stores already visited states to avoid re-exploration. The simulator keeps a hash map for this purpose, which is a common implementation technique to reduce the overhead involved in matching states. Execution states will only be explicitly compared, if their hash value is equal.

Normally only concrete state parts are considered for the computation of a hash value, since symbolic state parts can have many different representations with equal semantics. As already described in Section 6.1 the ESM method for state matching can also incorporate the symbolic state parts into the hash value computation. The resulting configuration will be denoted as ESM_H. This allows to filter out more necessarily unequal states without explicitly comparing

Table 7.2.: Evaluating symbolic state part hashing in a stateful search - first part

Benchmark	static POR			no POR	
	stateless	stateful		stateful	
	DFS	ESM_H	ESM	ESM_H	ESM
					Total execution time
term-counter-1.50	**1.494**	2.089	2.949	1.887	2.735
term-counter-1.100	**1.970**	3.089	6.588	2.801	6.331
term-counter-1.200	**3.045**	5.141	19.515	4.932	18.901
term-counter-1.400	**5.840**	10.491	67.150	10.180	66.846
term-counter-2.50	**2.109**	3.048	4.306	4.110	6.532
term-counter-2.100	**3.160**	4.839	10.212	7.616	17.367
term-counter-2.200	**5.884**	9.314	30.964	17.309	56.174
term-counter-2.400	**14.548**	21.972	108.884	47.385	201.798
term-counter-3.50	**2.975**	4.266	6.362	14.768	23.059
term-counter-3.100	**4.879**	7.187	15.102	28.466	56.607
term-counter-3.200	**10.331**	15.245	46.754	70.697	179.097
term-counter-3.400	**28.861**	39.778	165.537	215.655	T.O.
term-counter-4.50	**4.445**	6.306	8.985	136.086	320.093
term-counter-4.100	**7.543**	11.238	21.691	202.778	445.940
term-counter-4.200	**17.043**	24.240	66.046	412.263	T.O.
term-counter-4.400	**50.090**	65.849	233.793	T.O.	T.O.

Table 7.3.: Evaluating symbolic state part hashing in a stateful search - second part

Benchmark	static POR			no POR	
	stateless	stateful		stateful	
	DFS	ESM_H	ESM	ESM_H	ESM
					Number of hash matches
term-counter-1.50	N.A.	0	3580	1	3581
term-counter-1.100	N.A.	0	14655	1	14656
term-counter-1.200	N.A.	0	59305	1	59306
term-counter-1.400	N.A.	0	238605	1	238606
term-counter-2.50	N.A.	0	4774	79	8602
term-counter-2.100	N.A.	0	19524	129	34577
term-counter-2.200	N.A.	0	79024	229	139027
term-counter-2.400	N.A.	0	318024	429	557927
term-counter-3.50	N.A.	0	6027	1292	27596
term-counter-3.100	N.A.	0	24452	1542	91771
term-counter-3.200	N.A.	0	98802	2042	347621
term-counter-3.400	N.A.	0	397502	3042	T.O.
term-counter-4.50	N.A.	0	7523	53737	541429
term-counter-4.100	N.A.	0	29623	54587	699529
term-counter-4.200	N.A.	0	118823	56287	T.O.
term-counter-4.400	N.A.	0	477223	T.O.	T.O.

Table 7.4.: Evaluating symbolic state part hashing in a stateful search - third part

| Benchmark | static POR | | | no POR | |
| | stateless | stateful | | stateful | |
	DFS	ESM_H	ESM	ESM_H	ESM
				Number of explored transitions	
term-counter-1.50	**163**	163	163	165	165
term-counter-1.100	**313**	313	313	315	315
term-counter-1.200	**613**	613	613	615	615
term-counter-1.400	**1213**	1213	1213	1215	1215
term-counter-2.50	**233**	233	233	377	377
term-counter-2.100	**433**	433	433	677	677
term-counter-2.200	**833**	833	833	1277	1277
term-counter-2.400	**1633**	1633	1633	2477	2477
term-counter-3.50	**326**	326	326	1262	1262
term-counter-3.100	**576**	576	576	1962	1962
term-counter-3.200	**1076**	1076	1076	3362	3362
term-counter-3.400	**2076**	2076	2076	6162	T.O.
term-counter-4.50	**467**	467	467	7511	7511
term-counter-4.100	**767**	767	767	9211	9211
term-counter-4.200	**1367**	1367	1367	12611	T.O.
term-counter-4.400	**2567**	2567	2567	T.O.	T.O.

them. This section evaluates the performance benefits of doing so. A benchmark is particularly good suited for this case, if it satisfies the following requirements:

- Fast simulation to generate multiple states that can be matched

- Usage of simple expressions to avoid spending too much time solving complex constraints.

- Many different transition interleavings lead to states that have equal concrete state parts but different symbolic state parts.

The *term-counter* benchmark satisfies all of these requirements. It contains two (independent) threads that count up to a specified limit. A clock thread re-enables them in each time step. The number of threads and the count limit can be parameterized to create different benchmarks. Some results obtained by running this program with 1..4 threads and with 50, 100, 200 and 400 steps, are shown in Table 7.2, Table 7.3 and Table 7.4. The table shows the total execution times, the number of transitions explored and the number of hash matches between a currently explored state and one already visited. For each hash match, the states will be explicitly compared to determine whether they are equal.

A stateless POR search is an optimal configuration for this set of benchmarks, since it is sufficient to prune all redundant transition interleavings between the counter threads. Combining POR with a stateful search will not further reduce the explored state space, but add additional overhead. A stateful search will store visited states and compute hash values for them to ensure a more efficient lookup.

The POR with ESM_H variant performs second best and is not far away from the optimal stateless POR configuration. It shows an improvement of up to $x6.5$ compared to its ESM counterpart. Similar improvements can be observed between non partial order reduced ESM variants. Longer running simulation show greater improvements, due to the higher number

Table 7.5.: Comparing different SMT solvers

Benchmark	Z3 native	metaSMT		
		Z3	Boolector	Sword
condition-builder.8	19.292	4.800	**4.630**	4.672
condition-builder.16	T.O.	120.203	**114.909**	117.975
rbuf-2.1	2.294	1.922	1.950	**1.920**
rbuf-2.3	11.501	124.502	27.963	**11.119**
rbuf-2.5	49.052	392.463	38.417	**21.137**
symbolic-counter.1.3	2.672	1.808	**1.571**	1.575
symbolic-counter.1.6	4.232	2.167	1.981	**1.930**
symbolic-counter.1.9	5.903	2.859	2.653	**2.391**
symbolic-counter.1.12	7.740	3.188	2.717	**2.709**
symmetric-accumulator.3	5.546	11.097	**2.364**	6.438
symmetric-accumulator.4	10.298	24.325	**3.022**	107.718
symmetric-accumulator.5	21.569	52.845	**4.422**	T.O.
term-counter-1.50	3.795	2.119	**2.098**	2.133
term-counter-1.100	6.284	2.998	3.143	**2.988**
term-counter-1.200	11.378	**4.957**	5.496	5.075
term-counter-1.400	21.914	**9.048**	10.323	10.289

of hash matches which lead to explicit comparison of the states. Once the complexity of the analyzed program increases, the observed improvements diminish gradually, even though the number of hash matches increases. Apparently the simulation of transitions, which involves cloning of complex object structures, becomes more costly than the comparison of states.

Having only a single counter thread, thus POR will have no reduction effect, the stateful searches without POR finish slightly faster. The reason is that they do not apply a preliminary static analysis to obtain transition interference relations necessary for POR at runtime.

Result summary   The ESM method with symbolic state part hashing, denoted as ESM_H, shows the overall best results. Significant improvements can be observed compared to the standard ESM version because the number of spurious hash matches is greatly reduced.

7.4. Comparing SMT Solvers

The simulator supports a wide range of different solvers for SMT expressions by using the *metaSMT* back-end [Hae+11]. These include the *Z3* (version 4.1), *boolector* (version 1.5.118) and *sword* (version 1.1) SMT solvers. Unfortunately *metaSMT* currently does not provide the quantifier API supported by the Z3 solver. Thus the exact state subsumption method is not supported when using *metaSMT*.

For this reason direct support for the *Z3* solver through the provided Python bindings, without *metaSMT* in between, has been added. Using Z3 directly offers improved performance for benchmarks with involving more complex equality assumptions and supports quantifiers in SMT expressions. But it has been observed, that using Z3 directly often slows down the program simulation considerably, when compared to solvers accessed through *metaSMT*. The pre-built binary version *4.3.2* of Z3 is used.

The results of some selected benchmarks are shown in Table 7.5. The programs *term-counter* and *symbolic-counter* involve a lot of rather simple requests to the SMT solver. Whereas the *condition-builder* and *rbuf* programs require that more complex expressions are checked. The former involves non trivial logic expressions, whereas the latter involves some complex bitwise operations. A SPOR DFS with explicit structural matching (SPOR+ESM) is used as the base configuration.

The *boolector* solver accessed through *metaSMT* shows the overall best results, thus it will be used as the default solver in the following. The native Z3 solver can be considered second best. Since it is currently the only supported solver that provides support for quantifiers, it will also be used. The next section compares different state matching methods. Most of them involve the solver to check whether two symbolic state parts match (SBC-EQ/SBC-CV/ESS). Thus the employed solver has to handle a greater number of requests, which can also be more complex. Some configurations will use both the native Z3 solver and the *boolector* solver, therefore comparing them incidentally.

7.5. Comparing State Subsumption Matching Methods

This section provides a comparison of the different state matching methods. Basically they can be divided in explicit and solver-based methods. The former are more efficient (polynomial worst case complexity) but less precise than the latter.

The explicit methods are the Complete Equality Matching (CEQ), which requires that two states are completely equal, and the Explicit Structural Matching (ESM) with or without symbolic state part hashing, denoted as ESM_H or ESM respectively. The latter involves expression simplification rules, and can handle fresh symbolic literals compared to CEQ. Thus the CEQ method can be considered a special case of the ESM/ESM_H method. It can be used to show the benefits of the ESM method more clearly.

The solver-based methods are the heuristic (symbolic) equivalence (SBC-EQ), heuristic (symbolic) subsumption (SBC-CV) and exact (symbolic) subsumption (ESS). The heuristic methods will use the *boolector* solver through *metaSMT*. The configurations SBC-EQ+Z3 and SBC-CV+Z3 are similar but use the Z3 solver directly instead of the *boolector* solver through the *metaSMT* library. Thus this benchmark also incidentally compares the performance of these solvers. The ESS method is always used in combination with the direct Z3 solver, since it is currently the only supported solver that provides support for quantifiers, as required by the ESS method.

The total simulation times observed when applying the above configurations to a set of benchmarks are available in Table 7.6. The corresponding number of explored transitions are shown in Table 7.7. All of the programs considered in this benchmark set are safe. Most of them are based on the examples used in Chapter 6. In the following the benchmarks themselves will be briefly described and the observed results will first be interpreted and then evaluated. Finally some suggestions for future work are outlined based on the observed results.

7.5.1. Discussion of the Observations

The solver-based methods are in general able to prove all *rbuf* variants, those with and without time bound, since they are able to detect all cycles in the program. The ESM method on the other hand will miss all of them, since its expression simplification rules are insufficient to normalize the buffer operations, and thus the rotating buffer versions without time bound are not supported, since the simulation would not terminate. Specifying a time limit, allows to compare the runtime of the ESM and solver-based algorithms. As expected the solver-based methods perform clearly better, since they can match many redundant interleavings as already described in Section 6.2.1. Both heuristic solver-based methods perform equally on this benchmark, since

Table 7.6.: Comparison of different state subsumption methods, all employ SPOR

Benchmark	CEQ	ESM_H	SBC				ESS
			EQ	CV	EQ+Z3	CV+Z3	
condition-builder.8	N.S.	4.580	**2.071**	2.091	4.809	4.821	5.064
condition-builder.16	N.S.	112.111	**2.970**	2.989	9.276	9.011	9.696
condition-builder.32	N.S.	T.O.	**5.179**	5.233	18.165	18.510	19.593
counter-1.50	2.192	**1.901**	14.541	14.364	67.205	70.651	87.353
counter-1.200	**4.478**	4.685	T.O.	491.321	T.O.	T.O.	T.O.
counter-2.50	**2.491**	2.709	37.237	36.295	99.759	109.605	137.834
counter-2.200	**8.075**	8.437	T.O.	T.O.	T.O.	T.O.	T.O.
counter-3.50	**3.502**	3.818	75.000	73.043	146.316	161.465	198.165
counter-3.200	**13.146**	13.650	T.O.	T.O.	T.O.	T.O.	T.O.
counter-4.50	**5.226**	5.617	136.663	134.000	216.651	247.212	274.823
counter-4.200	**21.182**	21.590	T.O.	T.O.	T.O.	T.O.	T.O.
pc-sfifo-sym-1	T.O.	T.O.	T.O.	T.O.	T.O	T.O.	**1.877**
pc-sfifo-sym-2	T.O.	T.O.	T.O.	T.O.	T.O	T.O.	**2.166**
rbuf2-2	N.S.	N.S.	10.084	10.359	4.741	4.736	**3.238**
rbuf2-4	N.S.	N.S.	92.924	93.553	22.115	22.091	**7.177**
rbuf-2	N.S.	N.S.	26.486	26.375	5.362	5.446	**3.974**
rbuf-2.1	1.913	**1.890**	1.899	1.959	2.216	2.235	2.212
rbuf-2.3	26.828	27.784	25.936	26.490	5.223	**5.210**	5.219
rbuf-2.5	34.404	37.476	27.299	27.263	8.361	**8.297**	8.308
rbuf-2.9	T.O.	T.O.	32.966	33.062	14.629	14.355	**14.320**
rbuf-2.13	M.O	M.O	M.O	M.O	20.524	**20.414**	21.007
rbuf-4	N.S.	N.S.	77.491	78.410	28.779	29.365	**10.947**
rbuf-4.1	3.310	3.097	**2.680**	2.711	5.555	5.525	5.589
rbuf-4.3	T.O.	T.O.	32.105	32.114	**14.515**	14.635	15.018
rbuf-4.5	M.O	M.O	62.470	63.395	**23.595**	24.017	24.089
rbuf-4.9	M.O	M.O	M.O	M.O	**42.345**	42.751	42.757
rbuf-4.13	M.O	M.O	M.O	M.O	63.822	**61.847**	63.244
symbolic-counter.1.3	N.S.	1.887	1.934	**1.578**	5.227	3.279	1.941
symbolic-counter.1.6	N.S.	**1.921**	4.106	2.000	18.515	5.605	3.984
symbolic-counter.1.9	N.S.	**2.333**	7.850	2.560	41.552	9.185	7.411
symbolic-counter.1.12	N.S.	**2.696**	13.509	3.348	74.340	13.661	12.332
symbolic-counter.1.15	N.S.	**3.113**	21.145	4.423	114.903	19.879	18.811
symbolic-counter.2.9	N.S.	4.553	19.982	**2.886**	115.920	10.726	6.343
symbolic-counter.3.9	N.S.	7.067	28.489	**3.208**	165.473	12.316	5.285
symbolic-counter.4.9	N.S.	11.590	34.359	**3.642**	199.306	14.693	4.076
symbolic-counter.5.9	N.S.	18.577	36.316	4.087	205.478	16.745	**3.204**
symbolic-counter.6.9	N.S.	28.563	33.185	4.566	180.898	19.162	**2.533**
symbolic-counter.7.9	N.S.	39.224	25.803	5.139	133.851	21.238	**2.014**
symmetric-accumulator.1	**1.834**	1.840	T.O.	1.908	2.327	2.222	2.153
symmetric-accumulator.2	2.654	**1.976**	T.O.	2.076	2.853	2.661	T.O.
symmetric-accumulator.3	2.247	2.326	T.O.	**2.231**	3.451	3.116	T.O.
symmetric-accumulator.4	2.802	2.963	T.O.	**2.494**	3.976	3.608	T.O.
symmetric-accumulator.5	4.011	4.362	T.O.	**2.707**	4.599	4.122	T.O.
symmetric-accumulator.6	6.391	7.321	T.O.	**3.058**	5.241	4.657	T.O.
symmetric-accumulator.7	11.879	13.328	T.O.	**3.443**	5.930	5.268	T.O.
symmetric-accumulator.8	24.011	26.914	T.O.	**3.960**	6.693	5.988	T.O.
symmetric-accumulator.9	53.584	59.847	T.O.	**4.484**	7.541	6.695	T.O.
symmetric-accumulator.10	134.801	146.595	T.O.	**5.138**	8.477	7.641	T.O.
symmetric-accumulator.11	M.O	M.O	T.O.	**5.905**	9.730	8.652	T.O.
symmetric-accumulator.12	M.O	M.O	T.O.	**6.884**	10.881	9.743	T.O.
symmetric-accumulator.13	M.O	M.O	T.O.	**7.962**	12.282	11.042	T.O.
symmetric-accumulator.14	M.O	M.O	T.O.	**9.207**	13.873	12.475	T.O.
symmetric-accumulator.15	M.O	M.O	T.O.	**10.956**	16.195	14.638	T.O.
token-ring.1	N.S.	1.385	1.048	**1.018**	1.701	1.075	1.081
token-ring.6	N.S.	**1.564**	1.745	1.730	2.540	2.543	2.903
token-ring.9	N.S.	**2.089**	2.569	2.647	4.962	4.894	6.082
token-ring.13	N.S.	**4.108**	5.084	5.189	10.333	10.452	12.222
token-ring.15	N.S.	**5.424**	6.758	6.878	13.649	13.709	16.086
token-ring.17	N.S.	**7.030**	8.806	8.817	17.836	17.817	20.612
token-ring.20	N.S.	**9.763**	12.550	12.576	25.822	26.018	30.047
token-ring.25	N.S.	**15.470**	21.484	21.598	46.128	45.186	53.973

Table 7.7.: Number of explored transitions for Table 7.6

Benchmark	CEQ	ESM$_H$	SBC				ESS
			EQ	CV	EQ+Z3	CV+Z3	
condition-builder.8	N.S.	382	**64**	64	64	64	64
condition-builder.16	N.S.	12286	**136**	136	136	136	136
condition-builder.32	N.S.	T.O.	**280**	280	280	280	280
counter-1.50	159	159	159	159	159	159	**154**
counter-1.200	**609**	609	T.O.	609	T.O.	T.O.	T.O.
counter-2.50	225	225	225	225	225	225	**214**
counter-2.200	**825**	825	T.O.	T.O.	T.O.	T.O.	T.O.
counter-3.50	310	310	310	310	310	310	**286**
counter-3.200	**1060**	1060	T.O.	T.O.	T.O.	T.O.	T.O.
counter-4.50	435	435	435	435	435	435	**382**
counter-4.200	**1335**	1335	T.O.	T.O.	T.O.	T.O.	T.O.
pc-sfifo-sym-1	T.O.	T.O.	T.O.	T.O.	T.O	T.O.	**8**
pc-sfifo-sym-2	T.O.	T.O.	T.O.	T.O.	T.O	T.O.	**11**
rbuf2-2	N.S.	N.S.	12	12	12	12	**8**
rbuf2-4	N.S.	N.S.	30	30	30	30	**12**
rbuf-2	N.S.	N.S.	18	18	18	18	**12**
rbuf-2.1	10	10	**8**	8	8	8	8
rbuf-2.3	58	58	**20**	20	20	20	20
rbuf-2.5	250	250	**32**	32	32	32	32
rbuf-2.9	T.O.	T.O.	**56**	56	56	56	56
rbuf-2.13	M.O	M.O	M.O	M.O	**80**	80	80
rbuf-4	N.S.	N.S.	170	170	170	170	**68**
rbuf-4.1	114	114	**36**	36	36	36	36
rbuf-4.3	T.O.	T.O.	**104**	104	104	104	104
rbuf-4.5	M.O	M.O	**172**	172	172	172	172
rbuf-4.9	M.O	M.O	M.O	M.O	**308**	308	308
rbuf-4.13	M.O	M.O	M.O	M.O	**444**	444	444
symbolic-counter.1.3	N.S.	55	55	39	55	39	**21**
symbolic-counter.1.6	N.S.	109	109	57	109	57	**40**
symbolic-counter.1.9	N.S.	163	163	75	163	75	**58**
symbolic-counter.1.12	N.S.	217	217	93	217	93	**76**
symbolic-counter.1.15	N.S.	271	271	111	271	111	**94**
symbolic-counter.2.9	N.S.	425	284	87	284	87	**54**
symbolic-counter.3.9	N.S.	671	345	99	345	99	**48**
symbolic-counter.4.9	N.S.	1023	382	111	382	111	**41**
symbolic-counter.5.9	N.S.	1447	395	123	395	123	**35**
symbolic-counter.6.9	N.S.	1947	384	135	384	135	**29**
symbolic-counter.7.9	N.S.	2439	349	147	349	147	**23**
symmetric-accumulator.1	15	15	T.O.	**10**	10	10	10
symmetric-accumulator.2	37	37	T.O.	**17**	17	17	T.O.
symmetric-accumulator.3	81	81	T.O.	**24**	24	24	T.O.
symmetric-accumulator.4	169	169	T.O.	**31**	31	31	T.O.
symmetric-accumulator.5	345	345	T.O.	**38**	38	38	T.O.
symmetric-accumulator.6	697	697	T.O.	**45**	45	45	T.O.
symmetric-accumulator.7	1401	1401	T.O.	**52**	52	52	T.O.
symmetric-accumulator.8	2809	2809	T.O.	**59**	59	59	T.O.
symmetric-accumulator.9	5625	5625	T.O.	**66**	66	66	T.O.
symmetric-accumulator.10	11257	11257	T.O.	**73**	73	73	T.O.
symmetric-accumulator.11	M.O	M.O	T.O.	**80**	80	80	T.O.
symmetric-accumulator.12	M.O	M.O	T.O.	**87**	87	87	T.O.
symmetric-accumulator.13	M.O	M.O	T.O.	**94**	94	94	T.O.
symmetric-accumulator.14	M.O	M.O	T.O.	**101**	101	101	T.O.
symmetric-accumulator.15	M.O	M.O	T.O.	**108**	108	108	T.O.
token-ring.1	N.S.	**9**	9	9	9	9	9
token-ring.6	N.S.	**58**	58	58	58	58	58
token-ring.9	N.S.	**100**	100	100	100	100	100
token-ring.13	N.S.	**315**	315	315	315	315	315
token-ring.15	N.S.	**435**	435	435	435	435	435
token-ring.17	N.S.	**563**	563	563	563	563	563
token-ring.20	N.S.	**765**	765	765	765	765	765
token-ring.25	N.S.	**1125**	1125	1125	1125	1125	1125

the path condition is not relevant. The exact subsumption method can prove the *rbuf* program to be safe without simulating a complete cycle. It is sufficient to simulate a single step. The initial value of the bit buffer is symbolic. Rotating it once results again in a symbolic buffer which can assume all values that the original buffer could assume. All other state parts are equal. The *rbuf* benchmark performs some complex queries to the SMT solver to decide in each rotation step that the number of high bits in the buffer is not modified. This check itself seems to have a perceptible impact on the overall performance. The *boolector* solver is eventually unable to decide the queries, for benchmarks running over multiple time steps, within the given memory limit.

Another interesting result can be seen for the *symbolic-counter* example. It starts with a symbolic counter value that is between zero and a given bound. It will count up to a specified maximum value and then reset the counter back to its initial value[2]. It has already been described in Section 6.2.2, that once the counter is reset, the heuristic subsumption method is able to match it with the initial state, thus terminating early. The ESS method can match even more states in this example and thus prune the redundant state space more effectively.

The *pc-sfifo-sym-1/2* benchmarks also clearly show the advantage of the ESS method. Basically a consumer and producer communicate over a shared variable. They repeat their behaviour and increment an initially symbolic counter in each step. Essentially the heuristic methods fail to detect the subsumption between x and $x + 1$, where x is a symbolic value.

The *counter* program is similar to the *term-counter* program. It also has one to four independent threads that count up to a specified limit. But the clock thread, that notifies the counter threads in each step, will not terminate once all threads are done but keeps sending notifications. It is used as an example that can demonstrate the overhead involved in using a more sophisticated state matching technique when a simple one is already sufficient. Event the complete equality check is sufficient to verify this program. It can be observed that the explicit matching methods clearly show better results for this benchmark than the solver-based methods. Another example where the ESM method outperforms the solver-based methods is the *token-ring* benchmark.

Quite the opposite can be observed for the *condition-builder* benchmark. It involves the construction of logic expressions. The program runs without time bound and eventually starts to repeat its behaviour. The ESM method can than detect a cycle due to its logic expression simplification rules. But it is (currently) unable to match any redundant states where the logic expressions are reordered. The solver-based methods on the other hand can detect the equivalence due to the commutativity and prune much of the redundant search space.

The *symmetric-accumulator* program constructs a global result value over a specified number of time steps. In each step two dependent threads will be executed, thus both interleavings will be considered. But both of them generate a new semantically equivalent but structurally different symbolic expression and add it to the global result value. The simplification rules of the ESM are currently insufficient to detect the equivalence, thus it can be observed that the simulation time nearly doubles for each time step. The complete equality check (CEQ) method performs similarly. The heuristic solver-based methods are able to detect the equivalent states when the native Z3 solver is used but fail also when the *boolector* solver is used through *metaSMT*. The exact subsumption also fails since the native Z3 solver is unable to solve the resulting quantified formula.

---

[2]By subtracting the maximum value and adding the bound.

7.5.2. Result Summary

Altogether it can be observed that the complete equality (CEQ) method can be slightly faster than ESM but yields significantly weaker results. This is not an observation specific to this set of benchmarks but should also hold in general, since both algorithms are in principle equal, except that the ESM employs simplification rules and can handle the fresh symbolic literals problem. The measured slight performance gap between these two methods is because the CEQ method does not employ expressions simplification rules. But it can also have a negative impact on the performance, when large not simplified expressions are used as branch conditions or in assertions. Also the state matching can take more time. Thus the slight performance advantage for some benchmarks is clearly outweighed by the resulting disadvantages in general.

The results of the ESM method highly depend on the effectiveness of the expression simplification rules, since it does only match states which are equal up to the symbolic literals[3]. Thus adding additional simplification rules can improve the applicability of the ESM method to a larger set of benchmarks.

The heuristic solver-based subsumption method seems to be superior to the heuristic solver-based equivalence check. They are actually pretty similar except the subsumption method does not require the path conditions to be equal, but it is sufficient if one implies the other. This weaker state matching formulation seems not to pose a greater difficulty for the solver but allows to detect more state subsumptions, thus yielding a better state space reduction. In fact it can even be observed that the implication of path conditions is easier to check than equality, as the *symmetric-accumulator* benchmark suggests.

Apart from that there is no clear winner between the state matching methods. The ESM is considerably faster than the solver-based methods and still yields useful results. The exact subsumption method detects the largest number of state subsumptions. Its problem is that it can timeout if the state subsumption checking formulas become to complex. The heuristic solver-based methods are somewhere in between. They impose less additional overhead than the exact subsumption and can match more states than ESM. But still they also run slower than ESM, when exploring the same number of transitions, and can timeout when the subsumption checks become too complex.

Based on these results and observation it seems quite natural to combine different state subsumption methods. Some ideas for future work are sketched in the conclusion.

7.6. Comparing POR Implementations

This section compares the variants of the static and dynamic POR. Static persistent sets can currently be computed using the *conflicting transitions* (CT) or *stubborn set* (STUB) algorithm. The latter is available in two different configurations $STUB_1$, which will simply be called $STUB$, and $STUB_2$. It has already been described, that the *stubborn set* algorithm always yields better reductions than the *conflicting transitions* algorithm. The reason is that it always computes persistent sets (assuming the same input state and selected transition) that are a subset of those computed by the CT algorithm. This section compares both stubborn set algorithms and evaluates how much improvement they can yield compared to the CT algorithm.

Dynamic POR can optionally manage a *happens before* relation (HBR) at runtime to infer smaller persistent sets. When it is unable to compute non trivial persistent sets, DPOR can fall back to static persistent sets, instead of exploring all enabled transition in that state. Again

---

[3]Symbolic literals can be matched with other type compatible literals, when there is no contradiction, to handle the fresh symbolic literal problem as introduced in Section 6.3

either the CT or STUB algorithm can be used. The DPOR and its variants are compared against the SPOR.

The results of some benchmarks are available in Table 7.8. All configurations use a stateful search with the ESM method for state matching. First four configurations with SPOR are shown. They differ in the employed persistent set algorithm, as discussed above. The configuration shown in the fourth column computes all possible persistent sets for using the STUB algorithm and chooses the smallest one of them. Then three DPOR variants are shown. The first is the standard algorithm, the second manages a happens before relation and the last falls back to static persistent sets (using the STUB algorithm) when the inference of non trivial persistent sets fails dynamically.

Many programs yield the same results regardless which POR method is used, with regard to the number of explored transitions. Though the DPOR algorithms finish slightly faster, since they do not require a preliminary static analysis of the program. Some of the benchmarks show clear differences. It can be observed that the static STUB algorithm provides the best results compared with the other SPOR algorithms. Neither the STUB nor DPOR algorithm yield always better results than the other. The reason is that the STUB algorithm is more sophisticated, as it also considers which transitions can enable each other, but the DPOR algorithm has more precise informations at runtime. The best results for this benchmark set are obtained by combining static and dynamic POR. The last column shows the result of DPOR which falls back to the STUB algorithm whenever the inference of non trivial persistent sets fails.

The *buffer* program is an example where the DPOR excels. It first fills a buffer using multiple threads and then checks the result. All threads write to a different memory location, thus they are non-interfering. All different interleavings will eventually lead to the same resulting state. Detecting array index aliasing, which means whether $a[i]$ and $a[k]$ can overlap, statically is a complex problem in general, since i and k could be arbitrary (symbolic) values. The current static analysis simply assumes that every index access to the same array can possibly overlap. Consequently the SPOR analysis will explore all possible interleavings of the writer threads and thus run into the state explosion problem. Though in combination with a stateful exploration, the performance can be greatly increased. The DPOR exploration on the other hand will detect that the threads are non-interfering and only explore a single representative path.

Quite the opposite can be observed for the safe version of the *token-ring* and *transmitter* benchmarks. The DPOR algorithm often fails to infer non trivial persistent sets, since dependencies with disabled transitions are detected. SPOR in combination with a stubborn set algorithm performs much better, since it can still infer non trivial persistent sets in most cases.

Using a *happens before* relation (HBR) can indeed improve the results of the DPOR algorithm, though it does not yield any improvement for most of the considered benchmarks. Since maintaining this relation is a non trivial implementation task and can add additional runtime overhead, it is not necessarily recommended to use this option. In the following DPOR will be used without the HBR option.

Another interesting thing that can be observed from Table 7.8 is that always computing a locally minimal persistent set is not always an optimal decision, e.g. the unsafe *token-ring* benchmarks run better with the STUB algorithm choosing a single persistent set, than selecting the minimal one in each step. On the other hand it can still lead to an additional reduction as the *transmitter-safe* benchmark shows. Whether or not a minimal persistent set should be chosen depends among other on the additional overhead imposed by doing so. If the number of enabled transitions in a state is relatively small (and/or the explored states more complex, thus having proportionally more impact on the overall performance than the persistent set computation) the additional overhead may be negligible. In the following only a single persistent set will be

Table 7.8.: Comparing POR variants

Benchmark	V	ESM$_H$						
		static POR				dynamic POR		
		CT	STUB	STUB$_2$	STUB$_{MIN}$	standard	HBR	STUB
								total simulation time
kundu	S	3.314	3.335	3.322	3.379	**2.799**	2.908	3.382
mem-slave-tlm.5	S	4.777	4.771	4.789	4.745	**1.938**	1.939	4.822
pressure-safe.25	S	1.796	1.804	1.791	1.794	**1.718**	1.890	1.820
pressure-safe.100	S	4.048	4.090	4.119	4.124	**4.015**	7.429	4.103
simple-fifo-1c2p.20	S	5.230	5.286	5.213	5.186	**4.514**	4.573	5.317
simple-fifo-2c1p.20	S	7.143	7.126	7.163	7.143	**6.351**	6.442	7.267
simple-fifo-bug-1c2p.20	U	3.884	3.926	3.896	3.902	3.097	**3.082**	3.916
simple-fifo-bug-2c1p.20	U	4.245	4.262	4.252	4.261	3.449	**3.443**	4.263
simple-pipeline	S	28.103	3.247	3.236	3.295	**3.176**	3.209	3.277
symbolic-counter.1.15	S	3.227	3.152	3.181	3.159	**2.965**	3.174	3.230
symbolic-counter.7.9	S	40.704	**40.280**	40.318	40.345	41.204	45.384	41.129
symmetric-accumulator.5	S	4.574	4.487	4.483	4.485	**3.948**	4.030	4.591
symmetric-accumulator.8	S	27.905	**27.739**	27.800	27.814	27.989	29.349	28.902
term-counter-2.50	S	4.281	2.986	3.066	3.003	**2.743**	3.090	3.072
term-counter-3.100	S	26.083	7.411	7.405	7.307	**6.938**	10.119	7.450
term-counter-4.200	S	377.575	23.846	24.046	24.115	**23.720**	45.300	24.242
token-ring-bug2.9	U	53.363	3.920	53.734	3.795	**2.817**	2.899	4.019
token-ring-bug2.20	U	T.O.	7.839	T.O.	12.949	**5.206**	5.300	7.876
token-ring-bug.9	U	26.855	2.111	27.135	2.128	**1.638**	1.641	2.117
token-ring-bug.30	U	T.O.	5.670	T.O.	12.554	**3.777**	3.961	5.659
token-ring.9	S	57.833	2.143	58.761	**2.135**	30.778	39.270	2.184
token-ring.13	S	T.O.	4.229	T.O.	**3.180**	M.O	328.824	4.893
token-ring.20	S	T.O.	10.132	T.O.	**5.790**	T.O.	T.O.	12.910
transmitter-safe.9	S	42.354	1.462	1.528	1.628	25.634	**1.239**	1.505
transmitter-safe.13	S	T.O.	3.091	77.517	**2.284**	M.O	324.677	3.673
transmitter-safe.20	S	T.O.	7.310	250.949	**3.841**	T.O.	T.O.	9.654
transmitter.9	U	13.975	1.530	1.544	1.687	**1.201**	1.241	1.534
transmitter.13	U	359.549	1.882	38.390	2.309	**1.390**	1.432	1.880
transmitter.20	U	T.O.	2.452	124.449	3.892	**1.788**	1.882	2.530
								number of explored transitions
kundu	S	275	275	275	275	**263**	263	263
mem-slave-tlm.5	S	**27**	27	27	27	27	27	27
pressure-safe.25	S	**176**	176	176	176	176	176	176
pressure-safe.100	S	**701**	701	701	701	701	701	701
simple-fifo-1c2p.20	S	**31**	31	31	31	31	31	31
simple-fifo-2c1p.20	S	**43**	43	43	43	43	43	43
simple-fifo-bug-1c2p.20	U	**10**	10	10	10	10	10	10
simple-fifo-bug-2c1p.20	U	**10**	10	10	10	10	10	10
simple-pipeline	S	2317	**83**	83	87	129	129	83
symbolic-counter.1.15	S	278	**271**	278	271	271	271	271
symbolic-counter.7.9	S	2446	**2439**	2446	2439	2439	2439	2439
symmetric-accumulator.5	S	352	345	345	**344**	345	345	345
symmetric-accumulator.8	S	2816	2809	2809	**2808**	2809	2809	2809
term-counter-2.50	S	377	**233**	237	233	233	233	233
term-counter-3.100	S	1962	**576**	582	576	576	576	576
term-counter-4.200	S	12611	**1367**	1375	1367	1367	1367	1367
token-ring-bug2.9	U	5979	**170**	5979	170	170	170	170
token-ring-bug2.20	U	T.O.	**238**	T.O.	570	238	238	238
token-ring-bug.9	U	4145	**93**	4145	93	93	93	93
token-ring-bug.30	U	T.O.	**189**	T.O.	594	189	189	189
token-ring.9	S	9237	**100**	9237	100	3348	3320	100
token-ring.13	S	T.O.	315	T.O.	**170**	M.O	18413	315
token-ring.20	S	T.O.	765	T.O.	**331**	T.O.	T.O.	765
transmitter-safe.9	S	7427	**40**	40	65	3076	40	40
transmitter-safe.13	S	T.O.	237	10896	**113**	M.O	20083	237
transmitter-safe.20	S	T.O.	616	27464	**202**	T.O.	T.O.	616
transmitter.9	U	2335	**40**	40	65	40	40	40
transmitter.13	U	53291	**52**	5415	109	52	52	52
transmitter.20	U	T.O.	**73**	13685	199	73	73	73

computed.

Both the STUB and DPOR algorithms alone already yield good reductions. Their combination seems a natural choice if both algorithms have already been implemented. Based on these results it seems promising to investigate a more sophisticated DPOR implementation, that can compute non trivial persistent sets more often. It seems natural to employ the same idea that lead to the STUB algorithm. Whenever a dependency to a disabled transition is detected, only those transitions should be added, that can enable it. This set of transitions is called necessary enabling (NES). This set can either be obtained from a static analysis or somehow inferred dynamically at runtime. Both methods should yield a better reduction for DPOR than is obtained by falling back to STUB (which also pulls dependencies obtained by the static analysis and not only necessary enabling transitions). Further investigation of this approach is left for future research.

Result summary Neither SPOR nor DPOR is clearly superior. The combination of both of them shows the overall best results. Thus if both configurations are available it seems reasonable to combine them.

7.7. Comparing with Kratos

This section compares *complete symbolic simulation* approach developed in this thesis with the state of the art tool *Kratos*. The runtime results (in seconds) of the experiments are shown in Table 7.9 and Table 7.10. Due to page size constraints, some benchmarks are referred to by their short name, as defined in Table 7.1, and the runtime results are rounded to two decimal places. The column V denotes whether the program is safe (S) or contains bugs (U). All the configurations perform a partial order reduced search. The combined POR configuration is a combination of DPOR and SPOR, as already explained in the previous section. The static and combined POR use the stubborn set (STUB) algorithm to compute static persistent sets. A stateful search configuration can use different state subsumption algorithms, which have been evaluated in the previous sections. An exact method (ESS) and heuristic explicit- (ESM) and solver-based (SBC) methods. Either the Z3 solver is used directly (ESS and SBC-CV with Z3 column) or the *boolector* solver is used (all other configurations). Kratos is executed in the ESST configuration as specified in the *README* of the benchmark set for [CNR13] available from https://es-static.fbk.eu/people/roveri/tests/tcad-2012/:

```
./cswchecker --opt_cfa 2
            --inline_threaded_function 1
            --dfs_complete_tree=false
            --thread_expand=DFS
            --node_expand=BFS
            --po_reduce
            --po_reduce_sleep
            --arf_refinement=RestartForest
```

It can be observed that a stateful search clearly improves upon a stateless search. It supports the verification of cyclic state spaces and often is faster than the stateless DFS on acyclic state spaces, since it avoids redundant re-explorations. As already discussed in the previous comparison sections, the heuristic state matching methods can show significant improvements compared to the exact method (ESS). This effect can be observed particularly well for the ESM method. However the heuristics may fail to detect some valid subsumption, e.g. in the *pc-sfifo-sym-1/2* benchmark or *symbolic-counter*. The ESS configuration with SPOR is able to handle

Table 7.9.: Comparing with the state of the art tool Kratos

Benchmark	V	static POR					dynamic POR			combined POR		Kratos
		stateful					stateful			stateful		
		DFS	ESM$_H$	SBC-CV		ESS	ESS	ESM$_H$	DFS	ESS	ESM$_H$	ESST
				Z3								
bist-cell	S	2.25	2.24	2.31	2.54	2.53	1.53	**1.19**	1.20	2.55	2.22	1.29
buffer.p4	S	N.S.	2.28	2.65	5.74	5.46	2.05	**1.14**	N.S.	2.60	1.70	2.67
buffer.p5	S	N.S.	3.28	5.23	12.40	11.24	2.32	**1.20**	N.S.	2.89	1.84	5.14
buffer.p6	S	N.S.	5.57	13.16	30.27	25.58	2.56	**1.22**	N.S.	3.29	2.00	10.61
buffer.p7	S	N.S.	10.98	38.61	79.12	60.63	2.87	**1.35**	N.S.	3.65	2.14	23.80
buffer.p8	S	N.S.	23.67	114.55	210.02	146.20	3.10	**1.37**	N.S.	4.00	2.22	86.87
buffer.p9	S	N.S.	52.96	338.19	T.O.	342.19	3.41	**1.41**	N.S.	4.37	2.35	T.O.
buffer.p10	S	N.S.	119.48	T.O.	T.O.	T.O.	3.73	**1.49**	N.S.	4.74	2.51	T.O.
kundu	S	3.71	3.27	3.35	10.39	10.54	9.54	2.73	5.49	10.33	3.31	**1.05**
kundu-bug-1	U	1.42	1.46	1.43	1.57	1.56	1.20	1.02	1.03	1.57	1.44	**0.71**
kundu-bug-2	U	1.82	1.90	2.01	3.63	3.63	2.26	1.27	1.26	2.81	1.82	**0.58**
mst-bug2.1	U	4.11	4.13	4.24	4.88	4.94	2.33	1.59	1.57	4.82	4.15	**1.36**
mst-bug2.2	U	4.48	4.48	4.62	6.25	6.28	3.63	1.86	**1.82**	6.28	4.50	3.70
mst-bug2.3	U	4.85	4.94	5.23	9.15	9.26	6.54	2.27	**2.18**	9.16	5.14	11.08
mst-bug2.4	U	5.27	5.81	7.75	17.37	17.23	14.37	2.90	**2.79**	17.20	5.78	19.24
mst-bug2.5	U	6.13	7.53	18.98	45.09	42.73	39.29	4.46	**3.79**	41.92	7.62	37.42
mst-bug.1	U	3.89	3.84	3.81	4.47	4.49	2.25	**1.53**	1.54	4.49	3.81	2.41
mst-bug.2	U	4.00	4.08	4.10	5.37	5.33	2.96	**1.69**	1.69	5.26	4.11	9.59
mst-bug.3	U	4.26	4.24	4.22	5.30	5.31	2.81	1.67	**1.66**	5.26	4.19	29.77
mst-bug.4	U	4.56	4.59	4.66	7.24	7.21	4.60	**1.97**	1.98	7.17	4.60	87.34
mst-bug.5	U	4.75	4.93	4.94	8.29	8.16	5.54	**2.16**	2.16	8.27	4.91	188.09
mst.1	S	3.82	3.78	3.91	4.59	4.50	2.24	**1.51**	1.53	4.64	3.76	3.16
mst.2	S	3.97	3.97	4.06	5.33	5.37	2.95	**1.56**	1.67	5.31	4.01	12.45
mst.3	S	4.24	4.17	4.20	6.20	6.20	3.62	**1.69**	1.85	6.09	4.15	37.35
mst.4	S	4.44	4.41	4.39	7.02	7.00	4.42	**1.75**	2.00	7.06	4.38	308.52
mst.5	S	4.61	4.60	4.64	7.98	8.00	5.25	**1.92**	2.18	7.97	4.60	252.69
pc-sfifo-sym-1	S	N.S.	T.O.	T.O.	T.O.	1.85	1.53	T.O.	N.S.	1.85	T.O.	**0.46**
pc-sfifo-sym-2	S	N.S.	T.O.	T.O.	T.O.	2.10	1.69	T.O.	N.S.	2.15	T.O.	**0.60**
pres-safe.5	S	1.21	1.20	1.18	1.50	1.51	1.34	1.02	1.37	1.52	1.19	**0.40**
pres-safe.13	S	40.01	1.39	1.43	2.35	2.37	2.20	**1.23**	129.24	2.36	1.40	2.79
pres-safe.15	S	159.44	1.43	1.45	2.62	2.64	2.48	**1.27**	T.O.	2.58	1.51	4.58
pres-safe.20	S	T.O.	1.61	1.65	3.19	3.07	3.06	**1.47**	T.O.	3.16	1.62	7.83
pres-safe.25	S	T.O.	1.74	1.82	3.79	3.74	3.61	**1.62**	T.O.	3.68	1.77	17.73
pres-safe.50	S	T.O.	2.47	2.50	6.65	6.48	6.55	**2.38**	T.O.	6.49	2.51	205.77
pres-safe.100	S	T.O.	3.99	4.08	12.74	12.46	12.74	**3.95**	T.O.	12.67	4.05	T.O.
rbuf2-2	S	N.S.	N.S.	10.14	4.66	3.25	**2.98**	N.S.	N.S.	3.28	N.S.	54.56
rbuf2-3	S	N.S.	N.S.	93.24	22.06	7.17	**7.16**	N.S.	N.S.	7.17	N.S.	53.76
rbuf2-bug1.2	U	2.00	2.00	1.99	3.79	3.76	2.43	**1.36**	1.36	2.85	1.75	52.54
rbuf2-bug1.4	U	2.21	2.20	2.20	4.22	4.27	2.58	**1.40**	1.40	3.15	2.00	54.01
rbuf2-bug2.2	U	2.01	2.01	2.03	3.85	3.94	2.51	**1.35**	1.36	2.87	1.82	52.48
rbuf2-bug2.4	U	2.33	2.32	2.33	4.50	4.55	2.70	**1.50**	1.50	3.33	2.08	53.89
sfifo-1c2p.10	S	N.S.	3.48	5.60	24.90	23.80	23.52	**2.70**	N.S.	23.70	3.53	17.90
sfifo-1c2p.20	S	N.S.	5.09	16.30	54.90	45.65	44.98	**4.35**	N.S.	44.60	5.27	130.10
sfifo-1c2p.50	S	N.S.	11.99	142.17	232.06	114.68	113.58	**11.33**	N.S.	113.91	12.26	T.O.
sfifo-2c1p.10	S	N.S.	4.41	6.90	37.41	34.24	34.60	**3.58**	N.S.	35.95	4.45	18.09
sfifo-2c1p.20	S	N.S.	7.00	20.26	78.14	66.45	69.87	**6.26**	N.S.	67.38	6.95	110.60
sfifo-2c1p.50	S	N.S.	18.07	177.56	322.15	170.19	174.36	**17.47**	N.S.	174.29	18.09	T.O.

Continued on next page

Table 7.10.: Comparing with the state of the art tool Kratos– continued from previous page

| Benchmark | V | static POR | | | | dynamic POR | | combined POR | | | Kratos |
| | | DFS | ESM$_H$ | SBC-CV (Z3) | ESS | ESS | ESM$_H$ | DFS | ESS | ESM$_H$ | ESST |
			stateful			stateful			stateful		
sfifo-bug-1c2p.10	U	N.S.	2.81	3.03	9.17	9.23	8.46	**2.02**	N.S.	9.24	2.85	8.14
sfifo-bug-1c2p.20	U	N.S.	3.82	5.12	17.36	16.83	15.77	**3.05**	N.S.	16.99	3.78	64.95
sfifo-bug-2c1p.10	U	N.S.	2.96	3.21	10.40	10.24	9.37	**2.17**	N.S.	10.10	2.98	8.72
sfifo-bug-2c1p.20	U	N.S.	4.16	5.46	19.95	18.39	17.98	**3.33**	N.S.	18.72	4.12	51.31
simple-pipeline	S	T.O.	3.19	6.48	4.82	5.88	T.O.	**3.01**	T.O.	5.55	3.17	7.69
sym-cnt.1.3	S	N.S.	1.56	1.58	3.22	2.00	1.70	**1.33**	N.S.	1.98	1.59	9.24
sym-cnt.1.6	S	N.S.	1.90	2.01	5.60	3.96	3.67	**1.68**	N.S.	3.99	2.01	41.43
sym-cnt.1.9	S	N.S.	2.33	2.56	9.06	7.36	7.19	**2.08**	N.S.	7.28	2.35	175.43
sym-cnt.1.12	S	N.S.	2.71	3.37	13.59	12.34	12.12	**2.49**	N.S.	12.43	2.71	T.O.
sym-cnt.1.15	S	N.S.	3.03	4.37	19.46	18.71	18.56	**2.88**	N.S.	18.77	3.15	T.O.
sym-cnt.3.15	S	N.S.	12.72	**5.29**	23.85	14.99	14.83	12.74	N.S.	14.79	13.10	410.08
sym-cnt.6.15	S	N.S.	90.90	**7.10**	32.44	9.41	9.23	91.43	N.S.	9.46	91.27	T.O.
sym-cnt.9.15	S	N.S.	T.O.	9.62	43.16	5.23	**4.94**	T.O.	N.S.	5.24	T.O.	T.O.
token-ring-bug2.1	U	1.18	1.18	1.17	1.33	1.34	1.10	1.01	1.01	1.34	1.19	**0.21**
token-ring-bug2.13	U	3.71	5.19	36.72	42.93	6.95	5.42	3.47	3.05	6.93	5.19	**2.61**
token-ring-bug2.15	U	4.12	5.79	50.70	57.86	7.85	6.06	3.93	3.37	7.89	5.85	12.46
token-ring-bug2.17	U	4.58	6.50	70.09	75.46	8.76	6.77	4.35	3.78	8.88	6.50	M.O
token-ring-bug2.20	U	5.29	7.56	103.03	113.16	10.20	7.71	5.08	4.36	10.25	7.57	M.O
token-ring-bug.1	U	1.41	1.02	1.04	1.07	1.09	0.94	0.90	0.89	1.09	1.04	**0.15**
token-ring-bug.10	U	1.71	2.18	2.68	5.13	6.07	5.59	1.67	1.59	6.08	2.20	**0.54**
token-ring-bug.20	U	2.63	3.59	4.20	7.00	7.96	7.03	2.50	2.39	7.94	3.62	M.O
token-ring-bug.30	U	3.60	5.46	6.21	9.49	10.70	9.21	3.66	3.58	10.95	5.51	M.O
token-ring-bug.40	U	4.75	7.93	9.08	13.36	15.16	12.93	5.48	5.22	15.26	7.87	M.O
token-ring-bug.50	U	5.93	10.84	12.77	18.81	21.34	18.65	7.96	7.34	21.61	11.07	M.O
token-ring.1	S	N.S.	1.01	1.05	1.07	1.11	0.95	0.88	N.S.	1.08	1.04	**0.57**
token-ring.6	S	N.S.	1.56	1.83	2.61	2.86	8.79	2.98	N.S.	2.88	1.58	**0.59**
token-ring.9	S	N.S.	2.07	2.57	5.04	6.04	102.67	29.51	N.S.	6.06	2.13	**1.07**
token-ring.13	S	N.S.	5.45	5.12	10.43	12.14	T.O.	M.O	N.S.	12.75	**4.67**	5.22
token-ring.15	S	N.S.	7.58	6.82	13.68	15.82	T.O.	M.O	N.S.	17.16	**6.46**	148.90
token-ring.17	S	N.S.	**8.40**	8.82	17.81	20.57	T.O.	T.O.	N.S.	22.12	8.53	M.O
token-ring.20	S	N.S.	**9.78**	12.52	25.96	29.84	T.O.	T.O.	N.S.	32.76	12.51	M.O
token-ring.25	S	N.S.	**15.48**	21.56	45.66	54.26	T.O.	T.O.	N.S.	58.87	20.44	M.O
toy-bug-1	U	1.60	1.67	1.65	1.86	1.87	1.35	1.13	1.17	1.90	1.67	**0.57**
toy-bug-2	U	1.56	1.66	1.59	1.83	1.82	1.34	1.10	1.11	1.78	1.57	**0.27**
toy-sym	S	T.O.	T.O.	T.O.	T.O.	4.57	3.88	T.O.	T.O.	4.49	T.O.	**1.72**
transmitter.1	U	N.S.	0.94	0.94	1.01	1.03	0.91	0.89	N.S.	1.01	0.94	**0.14**
transmitter.10	U	N.S.	1.55	1.57	1.76	1.70	1.36	1.22	N.S.	1.73	1.54	**0.22**
transmitter.20	U	N.S.	2.48	2.41	2.65	2.62	1.93	1.75	N.S.	2.65	2.47	**0.29**
transmitter.30	U	N.S.	3.52	3.52	3.79	3.75	2.68	2.48	N.S.	3.72	3.62	**0.53**
transmitter.40	U	N.S.	4.79	4.84	5.09	5.02	3.61	**3.37**	N.S.	5.13	4.89	18.25
transmitter.50	U	N.S.	6.51	6.51	6.71	6.67	4.80	4.55	N.S.	6.72	6.58	**3.99**
transmitter.60	U	N.S.	8.27	8.32	8.59	8.42	6.13	**5.83**	N.S.	8.62	8.30	40.76
transmitter.70	U	N.S.	10.39	10.49	10.69	10.57	7.79	7.31	N.S.	10.87	10.54	**2.54**
transmitter.80	U	N.S.	12.62	12.64	12.98	12.79	9.47	9.18	N.S.	13.05	12.86	**3.61**
transmitter.90	U	N.S.	15.33	15.43	15.67	15.54	11.65	**11.48**	N.S.	15.90	15.55	M.O
transmitter.100	U	N.S.	18.30	18.64	19.08	18.69	14.85	**14.43**	N.S.	19.73	19.21	M.O

all chosen benchmarks within the time and memory limit. Combining SPOR and DPOR further improves the result, as already discussed in the previous section.

This combination shows very competitive results compared to Kratos. On some benchmarks improvements of several orders of magnitude can be observed. This can especially be observed with up-scaled benchmarks, e.g. the *token-ring-bug/2, transmitter, mem-slave-tlm* and *pressure* benchmarks. More interestingly the same result can be observed for many safe programs with cyclic state space, i.e. for the *token-ring, symbolic-counter, buffer* and *simple-fifo* benchmarks. Furthermore Kratos reports a wrong result (W.R.) for the *rbuf2* benchmarks. This might be explained with the complex symbolic bitvector operations that are employed in this benchmark.

8. Conclusion

A *complete symbolic simulation* approach has been proposed in this thesis for the verification of SystemC programs. This approach is built on top of SISSI [Le+13], to overcome its major limitation in verifying assertions over cyclic state spaces. Complete symbolic simulation is a combination of stateful model checking with symbolic simulation. The combined method allows to verify safety properties in (cyclic) finite state spaces, by exhaustive exploration of all possible inputs and process schedulings. Furthermore, stateful model checking avoids redundant re-exploration of already visited states, thus it can also improve the performance when applied to acyclic state spaces. The well-known state explosion problem is alleviated by integrating two complementary reduction techniques, namely Partial Order Reduction (POR) and State Subsumption Reduction (SSR):

- POR is a particularly effective technique to prune redundant scheduling interleavings. It works by exploring only a subset of enabled transitions in each state, that are provably sufficient to preserve all properties of interest. The sets of explored transitions can either be computed based on statically pre-computed or dynamically obtained dependency informations. The former approach is employed by SPOR, the latter by DPOR. Both variants are supported in this thesis in combination with a stateful search in the context of SystemC. The simultaneous combination of both techniques, which tries to infer dependencies dynamically and falls back to static information in case of failure, shows the best results.

  The *ignoring problem*, that can lead to unsoundness when naively combining POR with a stateful search, has been solved by incorporating a cycle proviso. The proviso used in this thesis conservatively ensures that for every state in the state space, obtained by applying POR, at least one fully expanded state is reachable.

  The stateful DPOR algorithm keeps a lightweight record of informations, which are relevant for the dependency analysis, for every explored transition to avoid the *missing dependency problem* during backtracking. For cyclic state spaces, additionally a novel approach has been proposed to update backtrack points conservatively, since the record of informations for states lying on a cycle may be incomplete.

- SSR can increase the effectiveness of symbolic state matching significantly. It is based on the observation, that symbolic states represent sets of concrete states. If the set of concrete states represented by a symbolic state s_1 is a subset of concrete states represented by a symbolic state s_2, then it is unnecessary to explore s_1, when s_2 already has been explored. The state s_1 is said to be *subsumed* by s_2 in this case. Subsumption is a generalization of equivalence, which already is difficult to detect in the context of symbolic execution. A basic stateful search can neither detect non-trivial equivalence nor subsumption between symbolic states, thus missing a lot of reduction opportunities. An exact algorithm, called ESS, has been proposed as part of SSR to detect subsumption by reducing the subsumption problem to the satisfiability of a quantified SMT formula. However, due to the use of quantifiers, this approach is computationally very expensive.

For this reason different heuristic methods have been proposed to detect subsumption more efficiently. They can improve the scalability, by spending less time in state matching, as it can often be faster to re-explore some (equivalent) part of the state space. Essentially two different heuristic approaches have been proposed: the *Explicit Structural Matching* (ESM) and the *Solver-Based Comparison* (SBC). They provide different tradeoffs between precision and runtime overhead. The ESM method attempts to structurally match all corresponding expression pairs. It offers polynomial worst case complexity but its precision depends on how well the expressions have been normalized in advance. The SBC method employs an SMT solver to achieve better precision at the cost of exponential worst case complexity. However, it still offers less runtime overhead than ESS, since no quantifiers are used.

The experimental evaluation reflects the anticipated results. The heuristics, especially the ESM, can result in huge improvements of the state matching efficiency. On the other hand the exact ESS algorithm can achieve much greater state space reduction. Both approaches can verify programs, where the other approach fails within a given time bound.

Significant scalability improvements have been observed when combining POR and SSR. This combination enables the efficient verification of non-trivial programs. The experiments show the competitiveness of the proposed *complete symbolic simulation* approach compared to Kratos, one of the state of the art model checkers for SystemC. Improvements of several orders of magnitude can be observed on some benchmarks. The results show that it is worthwhile to investigate further improvements for the current approach. Some ideas are described in the following.

8.1. Future Work

Complementary proof techniques Induction based proof methods can be employed to verify certain classes of programs, where a stateful search is inefficient (e.g. a concrete counter that will increase until an overflow). Another complementary optimization technique is the merging of execution paths. It is especially useful to alleviate the state explosion problem due to symbolic branches. However it has to be investigated how to efficiently combine it with a stateful search.

State caching and architectural improvements A stateful exploration normally has a significantly higher memory requirement than a stateless search, since (all) visited states are stored in a cache to avoid re-exploration. It might be worthwhile to investigate different caching strategies to decide which states to discard. This optimization is based on the observation that states not on the DFS search stack can be safely removed, while still preserving the correctness of the exploration algorithm, since they cannot form a cycle with the current search path. Another promising optimization is to share more data between states to allow for faster state cloning/backtracking and lower memory requirements. The employment of state collapsion methods, copy on write (COW) optimization and usage of persistent data structures can be investigated here.

Improving DPOR   The experiments show that neither DPOR nor SPOR is yields always better reductions compared to each other, despite the fact, that the DPOR algorithm has precise runtime informations available. Based on these results it seems promising to investigate a more sophisticated DPOR implementation, that can compute non trivial persistent sets more often. It seems natural to employ the same idea that lead to the stubborn set (STUB) algorithm, that is used to compute static persistent sets. Whenever a dependency to a disabled transition is detected, only those transitions should be added, that can enable it. This set of transitions is called necessary enabling (NES). This set can either be obtained from a static analysis, which is more lightweight than a complete dependency analysis required for SPOR, or somehow inferred dynamically at runtime, similarly to the dependency relation. Both methods should yield a better reduction for DPOR than is obtained by falling back to STUB, which will also pull dependencies obtained by the static analysis and not only the necessary enabling transitions. Further investigation of this approach is left for future research.

Combining different state subsumption methods   Based on the experiment results and observations it seems quite natural to combine different state subsumption methods. An idea would be to apply multiple state matching methods in parallel, e.g. start the ESM, the heuristic and exact subsumption checks and report a state match if either of these methods does match and else report that the considered states do not match. This might be especially useful on a multi-core machine. Since all these checks are read-only, they will not run into any synchronization issues. The difficulty here seems to be the cleanup. Whenever a method reports a positive result, the others should be terminated. Currently the SMT solver do not provide a reliable API to do so. A solution would be to fork new independent processes that do the actual work but that seems to much of an overhead.

Another useful direction that can be further investigated might be to apply the ESM first and only if it reports a negative result (no state match), than apply a solver-based method. The implementation is quite trivial, but some preliminary results obtained by doing so show no significant improvement compared to running the solver-based method alone. The reason seems to be that negative results are reported (much) more often than positive results. Thus the solver-based method is still called very often. The benefits obtained from some reduced calls to the solver are negated by the additional calls due to the ESM method. Its left for future research to investigate how to combine these methods effectively.

The *symmetric accumulator* benchmark is a good example that shows the exact subsumption check can be more performance intensive, than the heuristic checks. The exact subsumption is unable to infer that some states have already been visited. The checks are too complex and thus the time limit is exhausted. These experiments suggest that it might be useful to set a maximum time limit for symbolic state matchings in solver-based methods. Doing so could still detect more equivalent states than the ESM method, while not suffering too much from unsolvable instances, especially in acyclic state spaces.

This observation directly leads to another promising idea. The state matching method can be chosen dynamically at runtime for each pair of states (when their hash value and concrete state parts already match) s and v, where s is the currently explored state and v the already visited one. The rational behind this idea is that some state matchings are more important than others. States on a cycle are important. Failing to detect the cycle leads to non termination of the simulation. On the other hand if s and v do not form a cycle, s can be re-explored again. Detecting that s and v match is *merely* an optimization in this case.

A simple necessary condition for s and v to form a cycle is that v currently is in the search stack. So in this case a solver-based matching could be employed. Still it might be useful to start with a simple matching method and gradually increase the complexity. If s and v they really form a cycle, the states will be revisited until the cycle is detected (or the time/memory resources allocated for the simulation are exhausted). An idea based on this observation would be to keep a counter for each state on the stack, how often it has been (partially - considering the hash value and concrete state parts) matched. Based on the current value of the counter, a more sophisticated state matching algorithm can be employed or the time limit can be gradually increased.

If v is not on the search stack, it cannot form a cycle with s. So either a lightweight matching method, like the ESM, or a solver-based method with a relatively small timeout should be used. The timeout can either be a fixed value or itself dynamically chosen based on the expected time necessary to re-explore the state space of the already visited state. A simple method to approximate this time would be to store a timestamp for each state added to the stack. Whenever a state is backtracked another timestamp is recorded and the state is moved into the set of visited states[1]. The expected re-exploration time for the state is then the difference between the two timestamps.

---

[1]Or into a possibly size limited cache, as proposed in another point for future work.

A. Appendix

A.1. Generating Consistent Equality Assumptions

This section presents two algorithms, that can be used to generate consistent equality assumption sets. Every set of consistent equality assumptions can be used as argument to the formula $F_{\preccurlyeq}$ or $F_{\approx}$, defined in Definition 26 and Definition 27 respectively, to detect (symbolic) state subsumption or equivalence. The first algorithm is conceptually simpler and will generate all possible consistent sets, whereas the second algorithm will only generate a subset of them by filtering out sets of equality assumptions, that most likely will not lead to the detection of state subsumption. In the following the terms *(set of) equality assumptions* and *(symbolic literal) mapping* will be used interchangeably. First some common definitions and auxiliary functions, and then both algorithms will be presented.

A.1.1. Definitions

Indexing a set S by a number n, written as $S[n]$, returns any arbitrary element in S. The only requirement is, that the operation is deterministic, that is $S[n] = S[k]$ if $n = k$. Sometimes a set will be converted to a tuple, to make the indexing operation more explicit. Tuples will either be written in simple parentheses or brackets, e.g. $(1,2)$ or $[1,2]$. The auxiliary function *zip* takes two tuples A and B and yields a set S of pairs from A and B. It is defined as:

$$zip = \{(a,b) \mid a = A[i] \land b = B[i] \land i \in \{1..min(|A|,|B|)\}\}$$

For example $zip([1,2],[a,b,c]) = \{(1,a),(2,b)\}$. The operation $\Gamma(X)$ returns all symbolic literals reachable from X. It is defined for (symbolic) values, e.g. $\Gamma(x_1 + 2*x_2) = \{x_1,x_2\}$, and also for equality assumptions m as $\{e \mid (e = a \lor e = b) \land (a,b) \in m\}$, e.g. $\Gamma(\{(x_1,x_2),(x_3,y_1)\}) =$

Algorithm 16: PermutationsWithMixedRepetition(S, P, k)

Input: Sets S and P of arbitrary elements and a number k
Output: All k-permutations of S and P, where elements in S are selected without repetition and elements in P with repetition.

1   **function** *PermutationsWithMixedRepetition(S,P,k)***is**
2   **if** $k = 0$ **then**
3 | **return** $[[]]$
4    $ans \leftarrow []$
5   **for** $e \in (S \cup P)$ **do**
6 | **for** $R \in PermutationsWithMixedRepetition(S \setminus \{e\},P,k-1)$ **do**
7 | | $ans \leftarrow ans \cdot ([e] + +R)$
8   **return** *ans*

Algorithm 17: Collect all disjunct reachable symbolic literals from two states

Input: Structurally compatible states s_1 and s_2
Output: Pair of disjunct symbolic literal sets A and B reachable from s_1 and s_2 respectively

1 $A \leftarrow \{\}$
2 $B \leftarrow \{\}$
3 **for** $v_1, v_2 \in symbolicPairs(s_1, s_2)$ **do**
4 $\quad$ $A \leftarrow A \cup \Gamma(v_1)$
5 $\quad$ $B \leftarrow B \cup \Gamma(v_2)$

6 $C \leftarrow A \cap B$
7 $A \leftarrow A \setminus C$
8 $B \leftarrow B \setminus C$

Algorithm 18: Generate all sets of consistent equality assumptions between A and B.

Input: Disjunct sets A and B of symbolic literals
Output: Set of consistent equality assumption sets

1 **if** $|A| > |B|$ **then**
$\quad$ /* swap A and B, this ensures $|A| \leq |B|$ $\qquad\qquad\qquad\qquad$ */
2 $\quad$ $A, B \leftarrow B, A$
3 $M \leftarrow \{zip(tuple(A), P) \mid P \in PermutationsWithMixedRepetition(B, \{\varepsilon\}, |A|)\}$
4 **return** $\{m' \mid m \in M \wedge m' = \{(a, b) \in m \mid a \neq \varepsilon \wedge b \neq \varepsilon \wedge type(a) = type(b)\} \wedge m' \neq \emptyset\}$

$\{x_1, x_2, x_3, y_1\}$. Appending an element e to a tuple S is denoted as $S \cdot e$. Concatenating to tuples S and T is denoted as $S + +T$. Both operations are also defined on sets as $S \cup \{e\}$ and $S \cup T$ respectively.

An algorithm that can be used to compute k-permutations with mixed repetition is shown in Algorithm 16. It takes three arguments S, P and k. The number k denotes how many elements will be selected. Elements from S are selected without repetition. Elements from P are selected with repetition. For example let S=\{1,2,3\}, P=e and k=2. Then the result will be

$$[[1, 2], [1, 3], [1, e],$$
$$[2, 1], [2, 3], [2, e],$$
$$[3, 1], [3, 2], [3, e],$$
$$[e, 1], [e, 2], [e, 3], [e, e]]$$

Permutations with or without repetition can then be defined as:
PermutationsWithoutRepetition(S, k) = PermutationsWithMixedRepetition(S, \{\}, k)
PermutationsWithRepetition(S, k) $\quad$ = PermutationsWithMixedRepetition(\{\}, S, k)

A.1.2. First Algorithm

The complete algorithm consists of two parts. The first part is shown in Algorithm 17. Given two states s_1 and s_2. First all reachable symbolic literals from both states are collected in two separate sets A and B. Then all equal symbolic literals, those that are shared between both states, are removed from A and B, since they are implicitly equal and thus cannot be mapped upon another literal without resulting in inconsistent equality assumptions. Thus A and B contain all disjunct reachable symbolic literals from s_1 and s_2 respectively. Since both A and B are disjunct,

Algorithm 19: Preprocessing Phase

Input: Two structurally compatible execution states s_1 and s_2 ($s_1 \sim s_2$)
Output: Set of preprocessed (symbolic literal equality) constraints

1 **return** $generatePreprocessedConstraints(s_1, s_2)$

2 **function** $generatePreprocessedConstraints(s_1, s_2)$ **is**
3     $ans \leftarrow []$
4   **for** $v_1, v_2 \in symbolicPairs(s_1, s_2)$ **do**
5         $A \leftarrow \Gamma(v_1)$
6         $B \leftarrow \Gamma(v_2)$
7         $cs \leftarrow separateLiteralsBy((A,B), \lambda x : type(x))$
8   **for** $c \in cs$ **do**
9   **for** $e \in separateEqualLiterals(c)$ **do**
10                 $ans \leftarrow ans \cdot e$

11   **return** ans

12 **function** $separateLiteralsBy(c:(A,B), key)$ **is**
13   **return** $\{(a,b) \mid x \in A \cup B \wedge a = \{e \in A \mid key(e) = key(x)\} \wedge b = \{e \in B \mid key(e) = key(x)\}\}$

14 **function** $separateEqualLiterals(c:(A,B))$ **is**
15     $ans \leftarrow \{\}$
16     $common \leftarrow A \cap B$
17   **for** $e \in common$ **do**
18         $ans \leftarrow ans \cdot (\{e\}, \{e\})$

19     $ans \leftarrow ans \cdot (A \setminus common, B \setminus common)$
20   **return** ans

every symbolic literal $a \in A$ can be mapped upon any symbolic literal $b \in B$ or left unmapped. The second part is available in Algorithm 18. It shows how to generate all possible consistent equality assumptions from A and B. First it is ensured that $|A| \leq |B|$. Then all possible equality assumptions are generated in Line 3. Lastly all unmapped or type incompatible (to ensure consistency) pairs are filtered out of the set of equality assumptions resulting in all possible consistent equality assumption sets.

In the following, some remarks about the worst case complexity of this algorithm are presented. Assuming w.l.o.g. $|A| \leq |B|$, every $x \in A$ can be mapped upon a different $y \in B$ or can be left unmapped. So the first element $x_1 \in A$ can be associated with any element in B or left unmapped. The next $x_2 \in A$ has one less choice if x_1 has been mapped, else it will have the same amount of choices as x_1, and so on. Thus an upper bound for the number of generated mappings (equality assumption sets) is n^k. Where n is $max(len(A), len(B)) + 1$ and k is $min(len(A), len(B))$. A more precise upper bound is the number p of permutations returned from the function call $PermutationsWithMixedRepetition(B, \{\varepsilon\}, |A|)$ in line Line 3. If all symbolic literals in A and B are type combinations, that is $\forall a, b \in A \cup B : type(a) = type(b)$, then p is also a lower bound, since no mappings will be filtered out in Line 4.

The number of possible mappings grows very fast with the number of disjunct symbolic literals in s_1 and s_2. This algorithm is impractical for all but small such numbers. Additional filter heuristic, that discard mappings early that unlikely will lead to the detection of subsumption between states, are necessary, to make the algorithm scalable. An example heuristic would be to require that every $a \in A$ is mapped upon a $b \in B$ (assuming w.l.o.g. that $|A| \leq |B|$). This change can be easily integrated into the existing algorithm by changing the function call *PermutationsWithMixedRepetition*$(B, \{\varepsilon\}, |A|)$ into *PermutationsWithoutRepetition*$(B, |A|)$. Then an upper bound for the number of possible mappings would be $\frac{n!}{(n-k)!}$. If all symbolic literals in A and B are type compatible, then it would be exactly $\frac{n!}{(n-k)!}$.

A.1.3. Second Algorithm

The algorithm presented in this section generates only a subset of all possible equality assumption sets, by heuristically filtering out those equality assumptions, that will unlikely lead to the detection of subsumption between states. The algorithm consists of three subsequent phases: a preprocessing-, a construction- and a generation- phase.

The first two phases create an intermediate representation, namely a set of (equality) constraints. The generation phase then generates consistent equality assumptions from it. A constraint is a pair of symbolic literal sets (A,B). It denotes that any literal $a \in A$ can be mapped upon any literal $b \in B$ (or the other way round since equality is symmetric). Smaller constraints result in the generation of a smaller number of equality assumption sets. Constraints are reduced into smaller ones by splitting them into multiple disjunct parts. This process is denoted as separation. For example (A,B) can be separated into (A_1,B_1) and (A_2,B_2) where $A_1 \cap A_2 = \emptyset$ and $B_1 \cap B_2 = \emptyset$ [1]. The preprocessing phase constructs the initial constraints and separates each of them independently of the other ones yielding a set of preprocessed constraints. The construction phase than further separates the set of constraints by analyzing them all together yielding a simplified set of constraints. The following sections will first describe the preprocess-, construction- and generation- phase on after another. Then some additional heuristics and remarks will be presented.

Preprocessing Phase

The preprocessing phase takes two (structurally compatible) execution states s_1 and s_2 and generates a set of (preprocessed) constraints. The method is shown in Algorithm 19. The generated result set is denoted *ans* and is initially empty. Every symbolic state pair (v_1,v_2) between s_1 and s_2 is considered one after another. Each of them is transformed into an initial constraint $c = (A,B)$, where A and B are assigned all reachable symbolic literals from v_1 and v_2 respectively [2]. Next, in Line 7, the literals in A and B are separated by type. This operation yields a set of constraints cs, where $\forall c = (A,B) \in cs : (\forall a,b \in A \cup B : type(a) = type(b))$. Then every constraint $c \in cs$ is further separated by splitting equal symbolic literals, yielding constraints (A,B) where additionally $A \cap B \neq \emptyset$ iff $A = B \wedge |A| = 1$ holds. The resulting constraints are added to the preprocessed constraint set *ans*.

The following example shall illustrate the concepts of the presented algorithm. Given two states s_1 and s_2 with $s_1 \sim s_2$ and symbolic state parts:

$$\zeta(s_1) = (pc : (x_1 > 0) \wedge (x_3 \geq 5), \quad v : x_5 * (x_1 + x_2)^2, \quad w : x_3 + x_4 * x_6)$$

$$\zeta(s_2) = (pc : (y_1 > 0) \wedge (y_4 \geq 5), \quad v : y_5 * (y_1^2 + 2 * y_1 * y_2 + y_2^2), \quad w : x_6 * y_3 + y_4)$$

---

[1] In the following the properties $A_1 \cup A_2 = A$ and $B_1 \cup B_2 = B$ will also hold for each separated constraint, though it is not necessary in general.

[2] The reachable symbolic literals can be easily obtained, by a (recursive) structural walk over the corresponding value v_1, v_2.

Further it is assumed that x_5 and y_5 have type T_1. All other symbolic literals have the type T_2, with $T_1 \neq T_2$.

Initially the result set ans is empty. Both symbolic state parts have three entries, that can be grouped to pairs: the path conditions and the variable values v and w. First the pair $(pc(s_1), pc(s_2))$ is considered. Its initial constraint $c = (A, B)$ is computed as $A = \Gamma(pc(s_1)) = \{x_1, x_3\}$ and $B = \Gamma(pc(s_2)) = \{y_1, y_4\}$ in line 5 and 6. Since A and B neither contain any type incompatible nor equal literals, the constraint c is added to ans without further separation. Next the pair $(vars(s_1)[v], vars(s_2)[v])$ is considered. In this case $A = \{x_1, x_2, x_5\}$ and $B = \{y_1, y_2, y_5\}$. This constraint is separated by type into cs=$\{((\{x_1, x_2\}, \{y_1, y_2\}), (\{x_5\}, \{y_5\}))\}$ in line Line 7, since x_5 and y_5 have type T_1 and x_1, x_2, y_1, y_2 have a different type T_2. Neither of these two constraints $c_i = (A_i, B_i)$ is further separated in Line 9, since they don't contain any common symbolic literals, that is $A_i \cap B_i = \emptyset$. Both constrains are added to ans in Line 10. Lastly the pair $(vars(s_1)[w], vars(s_2)[w])$ is considered. In this case $A = \{x_3, x_4, x_6\}$ and $B = \{y_3, y_4, y_6\}$. This constraint is not separated by type, so $cs = \{(A, B)\}$ in Line 7. But it is separated by identity in Line 9 into two constraints $(\{x_3, x_4\}, \{y_3, y_4\})$ and $(\{x_6\}, \{x_6\})$, since $A \cap B = \{x_6\}$. Both are added to ans. The result of the preprocessing phase is then: ans = $\{((\{x_1, x_3\}, \{y_1, y_4\}), (\{x_1, x_2\}, \{y_1, y_2\}), (\{x_5\}, \{y_5\}), (\{x_3, x_4\}, \{y_3, y_4\}), (\{x_6\}, \{x_6\}))\}$.

Construction Phase

This section presents an algorithm, the construction phase, that takes a set of preprocessed constraints and transforms them into a set cs of simplified constraints. The algorithm is shown in Algorithm 20. The constraint set cs is simplified, if every pair of constraints $a, b \in cs$ is disjunct. A formal definition of this property is given in Line 27.

A constraint is just a pair of sets, so common set operations can be naturally extended to work with constraints. Doing so allows to express the following algorithm more naturally. Given two constraints $c_1 = (A_1, B_1)$ and $c_2 = (A_2, B_2)$:

$$c_1 \cap c_2 = (A_1 \cap A_2, B_1 \cap B_2)$$
$$c_1 \cup c_2 = (A_1 \cup A_2, B_1 \cup B_2)$$
$$c_1 \setminus c_2 = (A_1 \setminus A_2, B_1 \setminus B_2)$$
$$c_1 = \emptyset \iff A_1 = \emptyset \wedge B_1 = \emptyset$$
$$c_1 = c_2 \iff (A_1 = A_2) \wedge (B_1 = B_2)$$
$$size(c_1) = (|A_1|, |B_1|)$$
$$swap(c_1) = (B_1, A_1)$$

The construction phase starts with an empty result set cs. Each $c \in input$ is analyzed one after another using the function $tryAddConstraint$. This function expects that c is preprocessed (this property holds by definition of the preprocessing algorithm) and preserves the invariant that cs is simplified.

It starts by executing a loop where each constraint e currently in cs will be compared with c. Inside the loop first the common part $a = c \cap e$ of c and e will be computed. If a is empty or equals e, then e will be preserved (the if statement body in Line 9 will not be entered). Else, neither e and c are disjunct, nor e is fully contained in c, the constraint e will be separated into the common part a and the rest $b = e \setminus c$. So e is removed from cs and both a and b are added to cs. Doing so keeps cs simplified, since $a \cup b = e$ and e is disjunct to all other constraints in cs. Thus a and b are also disjunct to all other constraints in cs (which is the reason why they

Algorithm 20: Construction Phase

Input: Set of preprocessed constraints
Output: Set of simplified constraints

1 $cs \leftarrow \{\}$

2 **for** $c \in input$ **do**
3   ⌊ tryAddConstraint(c)

4 **@invariant**: isSimplified(cs)
5 **@require**: isPreprocessedConstraint(c)
6 **procedure** *tryAddConstraint(c : (A,B), first_step=True)* **is**
7   **for** $e \in cs$ **do**
8        $a \leftarrow e \cap c$
9   **if** $(a \neq \emptyset) \wedge (a \neq e)$ **then**
10          $cs \leftarrow cs \setminus e$
11   addNewConstraint(a)

12          $b \leftarrow e \setminus c$
13          $assert(b \neq \emptyset)$
14   addNewConstraint(b)

15        $c \leftarrow c \setminus e$
16   **if** $(c = \emptyset)$ **then**
17   ⌊ **return**

18   **if** *first_step* **then**
19   ⌊ tryAddConstraint(swap(c), False)

20   **else**
21   ⌊ addNewConstraint(c)

22 **@require**: $c \notin cs$
23 **procedure** *addNewConstraint(c=(A,B))* **is**
24   ⌊ $cs \leftarrow cs \cdot c$

25 **predicate** *isPreprocessedConstraint(c:(A,B))* **is**
26   ⌊ **return** $[(A \cap B \neq \emptyset) \iff (A = B \wedge |A| = 1)] \wedge (\forall a, b \in A \cup B : type(a) = type(b))$

27 **predicate** *isSimplified(cs)* **is**
28   ⌊ **return** $\neg$ *hasConstraintsWithCommonLiterals(cs)*

29 **predicate** *hasConstraintsWithCommonLiterals(cs)* **is**
30   ⌊ **return** $\exists (c_1, c_2) \in$ *2-combinations(cs)* $: \neg(c_1 \cap c_2 = \emptyset) \vee \neg(swap(c_1) \cap c_2 = \emptyset)$

31 **function** *2-combinations(cs)* **is**
32   ⌊ **return** $\{\{A, B\} \mid A \in cs \wedge B \in cs\}$

Table A.1.: Stepwise produced results for the example

step	considered $c \in input$	result set cs after this step
1	$(\{x_1,x_3\}, \{y_1,y_4\})$	$\{(\{x_1,x_3\}, \{y_1,y_4\})\}$
2	$(\{x_1,x_2\}, \{y_1,y_2\})$	$\{(\{x_1\}, \{y_1\}), (\{x_3\}, \{y_4\}), (\{x_2\}, \{y_2\})\}$
3	$(\{x_5\}, \{y_5\})$	$\{(\{x_1\}, \{y_1\}), (\{x_3\}, \{y_4\}), (\{x_2\}, \{y_2\}), (\{x_5\}, \{y_5\})\}$
4	$(\{x_3,x_4\}, \{y_3,y_4\})$	$\{(\{x_1\}, \{y_1\}), (\{x_3\}, \{y_4\}), (\{x_2\}, \{y_2\}), (\{x_5\}, \{y_5\}), (\{x_4\}, \{y_3\})\}$
5	$(\{x_6\}, \{x_6\})$	$\{(\{x_1\}, \{y_1\}), (\{x_3\}, \{y_4\}), (\{x_2\}, \{y_2\}), (\{x_5\}, \{y_5\}), (\{x_4\}, \{y_3\}), (\{x_6\}, \{x_6\})\}$

are added to cs without further analysis, since they are disjunct to all other constraints, they will not lead to any further separation).

Then in Line 15, c will be simplified to $c \leftarrow c \setminus e$. This is only an optimization that can make it unnecessary to traverse the whole current constraint list cs twice (since the function will call itself again in Line 19) every time a constraint is added. Doing so will not miss any separation of other constraints in cs, because cs is simplified, so common parts of c with e cannot be common parts with any other constraint $x \in cs$ where $x \neq e$. If c becomes empty, the function will return.

After the loop it is checked, whether this function has been called the first time for the currently analyzed constraint c. If it has been called the first time, then it will be called again with c being swapped. The reason is that it is also necessary to separate constraints $e = (A_2, B_2)$ with $c = (A_1, B_1)$ where $A_1 \cap B_2 \neq \emptyset$ or $B_1 \cap A_2 \neq \emptyset$. Else the function is called the second time. In this case c is not empty (else the analysis would have been already aborted in Line 17 or c would violate the property that it is preprocessed) and has no common parts with any constraint currently in cs (all common parts have already been separated). So c is added to cs. Doing so keeps cs simplified.

Remark. The above algorithm starts with an empty set of constraints $cs = \{\}$. This one is simplified by definition. Every constraint that is added to cs, this only happens in Line 24 and is initiated from the lines $\{11, 14, 21\}$, preserves the property, that cs is simplified. Thus cs is simplified when the algorithm finishes.

In the following, the construction phase algorithm will be applied to a small example for illustration. Given the set of preprocessed constraints (these are the result of the example used to illustrate the preprocess phase in the previous section):

$$input = \{(\{x_1,x_3\}, \{y_1,y_4\}), (\{x_1,x_2\}, \{y_1,y_2\}), (\{x_5\}, \{y_5\}), (\{x_3,x_4\}, \{y_3,y_4\}), (\{x_6\}, \{x_6\})\}$$

The construction phase starts with an empty result set cs. Each $c \in input$ is considered one after another. Since there are five $c \in input$, there will be five steps. The stepwise constructed result is available in Table A.1. The table shows the current step, the considered constraint $c \in input$ and the current result after analyzing c. First $c_1 = (\{x_1,x_3\}, \{y_1,y_4\})$ is directly added to cs, since cs is empty at the beginning. Next $c_2 = (\{x_1,x_2\}, \{y_1,y_2\})$ is considered. Since $c_1 \neq c_2$ and $a = c_1 \cap c_2 = (\{x_1\}, \{y_1\})$ is not empty (in Line 9), c_1 is removed from cs and the separated constraints a and $b = c_1 \setminus c_2 = (\{x_3\}, \{y_4\})$ are added instead. Then c_2 is simplified to $c_2' \leftarrow c_2 \setminus c_1 = (\{x_2\}, \{y_2\})$ in Line 15. Further separations do not occur, so c_2' is eventually added to cs (actually $swap(c_2')$ but it doesn't matter, since constraints are symmetric anyway). In step 3 the constraint $c_3 = (\{x_5\}, \{y_5\})$ is considered. Since neither $c_3 = (A, B)$ nor $swap(c_3) = (B, A)$ does overlap with any existing constraint currently available in cs, it is directly added to cs. Then in step 4 the constraint $c_4 = (\{x_3, x_4\}, \{y_3, y_4\})$ is considered. This one overlaps with the existing constraint $(\{x_3\}, \{y_4\})$ so it is separated into $(\{x_3\}, \{y_4\})$ and $(\{x_4\}, \{y_3\})$. The latter constraint is added to cs (the former is not, since it is already available). Finally in step 5 the constraint $(\{x_6\}, \{x_6\})$ is added to cs, similar to step 3, without any further separation.

Remark. Constraints of the form (A_1, B_1) with $A_1 = B_1$ are not filtered out prior the construction phase, because they can lead to the separation of constraints (A_2, B_2) if either $A_1 \cap A_2$, $A_1 \cap B_2$, $B_1 \cap A_2$ or $B_1 \cap B_2$ is not empty.

Generation Phase

Once a simplified set of constraints cs has been constructed, it can be used to generate sets of (simplified) consistent equality assumptions, as shown in Algorithm 21. Basically each $c = (A, B) \in cs$ is transformed into a set of locally (only for this single constraint) complete literal mappings. A globally (for all constraints in cs) complete literal mapping is obtained by choosing a mapping from each of the above local sets. Since cs is simplified, all of its constraints are disjunct, thus every such constructed mapping is consistent.

The following concrete example will be used to illustrate the algorithm, by showing the relevant intermediate results together with the end result. The following set of simplified constraints is given as input:

$$(a_1, a_2) \text{ - } (b_1, b_2)$$
$$(a_3, a_4) \text{ - } (b_3, b_4)$$
$$(a_5) \text{ - } (b_5)$$

Thus $cs[1] = (\{a_1, a_2\}, \{b_1, b_2\})$, and $cs[2] = (\{a_3, a_4\}, \{b_3, b_4\})$, and $cs[3] = (\{a_5\}, \{b_5\})$, and $cs = \{cs[1], cs[2], cs[3]\}$. Each constraint $c \in cs$ is transformed into a set of (local) equality assumption sets in Line 8. The result of each transformation is denoted as M. Since three constraints $cs[i]$ are available, three corresponding sets M_i will be produced, with:

$$M_1 = \{\{(a_1, b_1), (a_2, b_2)\}, \{(a_2, b_1), (a_1, b_2)\}\}$$
$$M_2 = \{\{(a_3, b_3), (a_4, b_4)\}, \{(a_4, b_3), (a_3, b_4)\}\}$$
$$M_3 = \{\{(a_5, b_5)\}\}$$

Neither of them is empty, thus they are all collected as $choices = [M_1, M_2, M_3]$. The next step is to combine all $e_i \in M_i$ with each other. This happens by computing the cartesian product of $[M_1, M_2, M_3]$ in Line 11, which results in:

$$\begin{aligned}
product([M_1, M_2, M_3]) = \{&(\{(a_1, b_1), (a_2, b_2)\}, \{(a_3, b_3), (a_4, b_4)\}, \{(a_5, b_5)\}), \\
&(\{(a_1, b_1), (a_2, b_2)\}, \{(a_4, b_3), (a_3, b_4)\}, \{(a_5, b_5)\}), \\
&(\{(a_2, b_1), (a_1, b_2)\}, \{(a_3, b_3), (a_4, b_4)\}, \{(a_5, b_5)\}), \\
&(\{(a_2, b_1), (a_1, b_2)\}, \{(a_4, b_3), (a_3, b_4)\}, \{(a_5, b_5)\})\}
\end{aligned}$$

The last step is to flatten all inner sets into single top level sets, which results in a set of four different consistent equality assumption sets:

$$\begin{aligned}
flatten(product([M_1, M_2, M_3])) = \{&\{(a_1, b_1), (a_2, b_2), (a_3, b_3), (a_4, b_4), (a_5, b_5)\}, \\
&\{(a_1, b_1), (a_2, b_2), (a_4, b_3), (a_3, b_4), (a_5, b_5)\}, \\
&\{(a_2, b_1), (a_1, b_2), (a_3, b_3), (a_4, b_4), (a_5, b_5)\}, \\
&\{(a_2, b_1), (a_1, b_2), (a_4, b_3), (a_3, b_4), (a_5, b_5)\}\}
\end{aligned}$$

Algorithm 21: Generation Phase

Input: Set of (simplified) constraints
Output: Set of consistent equality assumption sets

1 **function** $product(S = (S_1,...,S_N))$ **is**
2    $\lfloor$ **return** $\{(a_1,...,a_n) \mid \bigwedge_{i \in \{1..n\}} a_i \in S_i\}$

3 **function** $flatten(S)$ **is**
4    $\lfloor$ **return** $\{\bigcup_{e \in X} e \mid X \in S\}$

5 **for** $c = (A,B) \in constraints$ **do**
6   **if** $|A| > |B|$ **then**
      /* swap A and B, this ensures $|A| \geq |B|$ to compute the following
      permutations correctly   */
7      $\lfloor$ $A, B \leftarrow B, A$
8    $M \leftarrow \{zip(tuple(A),P) \mid P \in PermutationsWithoutRepetition(B,|A|)\}$
9   **if** $M \neq \emptyset$ **then**
10      $\lfloor$ $choices \leftarrow choices \cdot M$

11 **return** $flatten(product(choices))$

The number of equality assumption sets that will be generated by the above algorithm is: $\prod_{i=1}^{|cs'|} \frac{n_i!}{(n_i-k_i)!}$ with $cs' = \{(A,B) \in cs \mid A \neq \emptyset \wedge B \neq \emptyset\}$ and $(A,B) = cs'[i]$ and $n_i = max(|A|,|B|)$ and $k_i = abs(|A| - |B|)$. The inner term $\frac{n_i!}{(n_i-k_i)!}$ is due to the permutations in Line 8. The outer product corresponds to the cartesian product in Line 11. Filtering out constraints that will not generate any equality assumptions happens in Line 9.

The number of equality assumption sets generated by this method does no longer grow exponentially with the number of all disjunct symbolic literals, but with the size of the maximum constraint $c = (as,bs) \in cs$, which is defined as $max(|as|,|bs|)$. An upper bound for this maximum size is the maximum number of disjunct symbolic literals for each linked symbolic pair:

$$max(\{max(|\Gamma(v_1) \setminus \Gamma(v_2)|, |\Gamma(v_2) \setminus \Gamma(v_1)|) \mid (v_1,v_2) \in \texttt{symbolicPairs}(s_1,s_2)\})$$

The preprocessing phase starts with such maximal constraints for each pair (v_1,v_2), which is in $\texttt{symbolicPairs}(s_1, s_2)$. These constraints can only be separated into smaller ones (using simplification heuristics), but they will never grow.

So this algorithm to generate equality assumptions has still an exponential worst case complexity. But in practice it might often perform (considerably) better.

Example 4. The example from the preprocessing and construction phase started with the symbolic state parts:

$$\zeta(s_1) = (pc: (x_1 > 0) \wedge (x_3 \geq 5), \quad v: x_5 * (x_1 + x_2)^2, \quad w: x_3 + x_4 * x_6)$$
$$\zeta(s_2) = (pc: (y_1 > 0) \wedge (y_4 \geq 5), \quad v: y_5 * (y_1^2 + 2 * y_1 * y_2 + y_2^2), \quad w: x_6 * y_3 + y_4)$$

And resulted in the simplified constraint set:

$$\{(\{x_1\}, \{y_1\}), (\{x_3\}, \{y_4\}), (\{x_2\}, \{y_2\}), (\{x_5\}, \{y_5\}), (\{x_4\}, \{y_3\}), (\{x_6\}, \{x_6\})\}$$

The generation phase would then yield only a single set of equality assumptions, namely $\{(x_1,y_1), (x_3,y_4), (x_2,y_2), (x_5,y_5), (x_4,y_3)\}$. This one is already sufficient to show that s_1

and s_2 are equivalent. And it even seems to be necessary (other consistent equality assumptions would not lead to the detection of s_1 and s_2 being equivalent). The first algorithm (presented in Section A.1.2) on the other hand would generate either 1546 or $\frac{5!}{(5-5)!} = 5! = 120$ sets of equality assumptions, depending whether all- or only (locally) complete- sets are generated. So the second algorithm can lead to significant performance improvements.

Additional Heuristics

The preprocessing phase of the algorithm separates all initial constraints $c = (A, B)$ by type. This is required so that the generation phase will yield type-compatible equality assumptions, which is a necessary condition for them to be consistent. Additional separation criteria can be used to generated smaller constraints. It seems reasonable to separate constraints by source(-file) location of the symbolic literals. Doing so can often yield (considerably) smaller constraints but should still be sufficient to detect all state cycles (since the same statements would be executed for each cycle, therefore all introduced symbolic literals should have the same source locations as the ones introduced on the previous cycle execution). Two literals that have the same source location necessarily have the same type, thus it preserves type-compatibility. So separation by location can be used as drop-in replacement to type based separation (the key function $\lambda x : type(x)$ would just be replaced by $\lambda x : location(x)$ in Line 7 of the preprocessing Algorithm 19). It also implies that constraints separated by location cannot be larger than those separated by type.

The construction phase accepts any new constraint $c = (A, B)$ into the collected constraint set cs so long as cs stays simplified by adding c. This means, by definition of simplified, that neither two different constraints in cs have any common symbolic literals. Simplification is ensured, by comparing the new about to be added constraint with all existing constraints, further separating them if necessary. Ultimately all these separated constraints will be accepted, since they keep the constraint set cs simplified.

Additional requirements can be defined to decide whether a new constraint is accepted or rejected. When a constraint is rejected, the whole algorithm aborts and the compared states s_1 and s_2 are considered to be non-equivalent. The idea is to require that $|A| = |B|$ holds, as it seems rather uncommon that constraints with $|A| \neq |B|$ will lead to equality assumptions that in turn lead to the detection of equivalence (or more generally state subsumption). Because if $|A| \neq |B|$ then at least one symbolic literal will be left unmapped (actually $abs(|A| - |B|)$). So basically If $|A| = |B|$, than the constraint $c = (A, B)$ is called *regular*, else *irregular*. The special case where $abs(()|A| - |B|) \leq 1$ will be called *almost regular*.

The following example may illustrate the concept more clearly. Given two symbolic state parts $\zeta(s_1)=(pc:\ T,\ a:\ x_1 * x_2 + x_1,\ b:\ x_3)$ and $\zeta(s_2)=(pc:\ T,\ a:\ y_1 * (y_2 + 1),\ b:\ y_1)$. Further assuming that all symbolic literals x_1, x_2, x_3, y_1, y_2 have the same type and constraints are separated by type, the preprocessing phase will yield the two constraints $c_1 = (\{x_1, x_2\}, \{y_1, y_2\})$ and $c_2 = (\{x_3\}, \{y_1\})$. The construction phase starts with an empty constraint set cs. In the first step c_1 is added to cs. Next c_2 is considered. Since $c_1 \cap c_2 = (\{\}, \{b_1\})$ is not empty, both constraints will be further separated, resulting in $cs=\{(\{a_1, a_2\}, \{b_2\}), (\{\}, \{b_1\}), (\{a_3\}, \{\})\}$ If regular constraints are required, then the construction phase and with it the whole algorithm would abort without generating any equality assumptions. The states s_1 and s_2 would be considered non-equivalent without calling the solver at all. Else two different simplified consistent equality assumption sets, namely $m_1=\{(a_1, b_1)\}$ and $m_2=\{(a_2, b_1)\}$ will be generated. So the solver would be called twice to detect, that s_1 and s_2 are non-equivalent.

Requiring regular constraints is additional heuristic, as for example $\zeta(s_1)=(pc:\ T,\ a:\ 5)$ and

$\zeta(s_2)=(pc\colon x_1 = 5, a\colon x_1)$ would not be detected as being equivalent, since the only generated constraint $(\{\}, \{x_1\})$ is not regular. Though many such cases, including this one, could already be simplified in advance, e.g. replacing x_1 with 5 since $x_1 = 5 \in pc(s_2)$ anyway. Or the requirement that constraints must be regular could be relaxed to requiring almost regular constraints (or more generally $abs(|A| - |B|) \leq k$).

Remark. Preprocessing and the subsequent construction don't have to be completely separate phases. They can easily be intertwined by just directly calling *tryAddConstraint* from Algorithm 19 in Line 10, instead of collecting the constraint. Doing so can be useful, if heuristics are employed, that allow to abort early, like requiring regular constraints does. This can often make it unnecessary to generate all preprocessed constraints, when only some of them would be sufficient to abort.

Remark. Other algorithms can be used to generate equality assumptions m which can than be used to detect subsumption between states. The only requirement is that m is consistent (type-compatible and unambiguous).

Remarks

As already stated, this second algorithm generates only a subset of all possible equality assumptions. It filters out those, that will only unlikely lead to the detection of state subsumptions. More precisely the algorithm generates equality assumptions that satisfy the following definition.

Definition 29

> Let m be a set of equality assumptions for the states s_1 and s_2, m is called simplified *iff the following properties C_1 and C_2 hold (with $P = \text{symbolicPairs}(s_1, s_2)$)*
>
> $C_1 : \ \forall (v_1, v_2) \in P : (\forall a \in \Gamma(v_1) : a \in \Gamma(m)) \vee (\forall b \in \Gamma(v_2) : b \in \Gamma(m))$
>
> $C_2 : \ \neg\exists(a,b) \in m : (\exists(v_1, v_2) \in P : (x, y \in \{a, b\} \wedge \neg(x \in \Gamma(v_1) \iff y \in \Gamma(v_2))))$
>
> *A set of equality assumptions that satisfies C_1 will be called* (locally) complete.

Basically C_1 states that for every symbolic state pair (v_1, v_2) either all symbolic literals from v_1 have been mapped or all symbolic literals from v_2 have been mapped. If only regular constraints are accepted in the construction phase, then every symbolic literal of v_1 and v_2 is mapped (there exist no unmapped literal).

The property C_2 states that every equality assumption $(a,b) \in m$ must appear in corresponding symbolic state pairs. So for every symbolic state pair (v_1, v_2) if a appears in v_1 (or v_2) then b must appear in v_2 (or respectively v_1).

A.2. POR Correctness Proof: Deadlock Preserving

This sections provides a proof for the following theorem introduced in Section 3.1.1 as Theorem 3.1.1.

Theorem A.2.1 (*Deadlock preserving partial order reduction*)

> Let A_R be a reduced state space where the reduction function r satisfies the conditions C_0 and C_1. Let s_d be a deadlock reachable from the initial state s_0 in A_G by a trace w. Then s_d is also reachable from s_0 in A_R.

A similar proof has already been provided in [God96]. The proof in this section is slightly differently formulated and adapted to the terminology used in this thesis. It is provided here to serve as an introduction to more advanced proofs.

First the overall proof method is outlined. The idea is to show that for every state s in A_R and every trace w that leads to a deadlock in A_G from s, there exists an equivalent trace $w_i \in [w]_s$ that leads to a deadlock in A_R from s. The Definition 6 of equivalent traces implies that w and all w_i have the same length. The proof proceeds by induction. The induction step requires some auxiliary lemmata which are introduced in the following. Basically it is shown that at least one t_i, which is the first transition of w_i, is in the persistent set of s. Thus some progress is made toward the deadlock in A_R. Together with the induction hypothesis the theorem can be finally proved. In the following first the auxiliary lemmata are introduced then the main theorem will be proved as outlined above.

Lemma A.2.2. *Let s be a state in A_R. Let d be a deadlock reachable in A_G from s by a non empty trace w. Let T be a non empty persistent set in s. Then at least one of the transitions in the trace w is in T.*

Proof. Let $w = t_1..t_n$ and let $s = s_1 \xrightarrow{t_1} s_2 \xrightarrow{t_2} ...s_n \xrightarrow{t_n} s_{n+1} = s_d$ be the sequence of states w goes through in A_G. The proof proceeds by contradiction. Assume that neither of the transitions t_j, with $j \in \{1..n\}$, is in the persistent set T of s. Then by Definition 10 of persistent sets, all transitions $t \in T$ are independent with t_j in s_j for $j \in \{1..n\}$. Thus by Definition 4 of transition independence all transitions in T remain enabled in all s_j and also in s_d. By assumption T is not empty. Hence s_d cannot be a deadlock. Consequently, at least one of the transition in w must be in the persistent set T of s. $\square$

Lemma A.2.3. *Let s be a state in A_R. Let d be a deadlock reachable in A_G from s by a non empty trace w. For all $w_i \in [w]_s$ let a_i denote the first transition of w_i. Let T be a non empty persistent set in s. Then at least one of the transitions a_i is in T.*

Proof. The proof proceeds by contradiction. Assume that neither a_i is in T. Let $w = t_1..t_n$ and let $s = s_1 \xrightarrow{t_1} s_2 \xrightarrow{t_2} ...s_n \xrightarrow{t_n} s_{n+1} = s_d$ be the sequence of states w goes through in A_G. By Theorem A.2.2 at least one transition of w is in T. Let t_k be the first such transition ($k \le j$ for all t_j of w that are in T). According to Definition 10 of persistent sets, t_k is independent with all t_j in s_j for $j \in \{1..k-1\}$. Thus by Definition 6 of trace equivalence, the traces $w' = t_k t_1..t_{k-1} t_{k+1}..t_n$ and $w = t_1..t_n$ are equivalent, so $w' \in [w]_s{}^3$. Since t_k is the first transition of w' and t_k is in T, the proof is completed.

$\square$

Theorem A.2.4

> Let s be a state in a reduced STS A_R, where the reduction function r satisfies the conditions C_0 and C_1 as defined in Section 3.1.1. Let s_d be a deadlock reachable from s in A_G by a trace w. Then s_d is also reachable from s in A_R.

Proof. The proof proceeds by induction over the length of the trace w.

IB : $|w| = 0$. This case immediately holds, since executing an empty trace does not change the state (thus $s = s_d$).

---

[3]And analogously $w \in [w']_s$ but this result is irrelevant for the proof.

IS : $|w| = n + 1$

> Since a non empty trace is executable from s in A_G, $enabled(s)$ cannot be empty. Consequently $r(s)$ is not empty due to condition C_0. Condition C_1 ensures that $r(s)$ is a persistent set in s.
>
> Thus, according to Theorem A.2.3 there exists a trace w_i in A_G such that $w_i \in [w]_s$ and the first transition t_i in w_i is in $r(s)$, hence t_i is also explored in A_R. Executing t_i from s results in a state s' in A_R (since t_i is explored in A_R from s). Since $w_i \in [w]_s$, executing either w_i or w from s will result in the same state, according to Definition 2.6.1 of trace equivalence. Thus the deadlock state s_d is reached from s when executing w_i in A_G, i.e. $s \xrightarrow{w_i} s_d$. Let w' be the rest of w_i (all transitions except t_i).
>
> So s' is a state in A_R, whose reduction function r by assumption satisfies the conditions C_0 and C_1, and the deadlock s_d is reachable from s' in A_G by the trace w'. Consequently, by induction hypothesis (IH) the deadlock s_d is also reachable from s' in A_R.

$\square$

From the above theorem it follows immediately that Theorem A.2.1 holds, since the initial state s_0 is in A_R. Thus the reduced state space preserves all deadlocks of the corresponding complete state space A_G.

A.3. POR Correctness Proof: Assertion Violations Preserving

This sections provides a proof for the theorem introduced in Section 3.1.3 as Theorem 3.1.3.

Theorem A.3.1

> Let A_R be a reduced state space where the reduction function r satisfies the conditions C_1 and C_2 as defined in Section 3.1. Let w be a trace in A_G that leads to an error state from the initial state s_0. Then there exists a trace w_r in A_R from s_0 such that $w \in Pref([w_r]_{s_0})$.

The expression $Pref([w]_s)$ denotes the set of prefixes of the sequences in $[w]_s$ for every trace w and state s (thus $[w]_s \subseteq Pref([w]_s)$ always is valid), as described in Section 3.1.3.

According to [EP10] it can be shown, though no proof is available there, that a reduced state space A_R, corresponding to a complete state space A_G, where the reduction function r satisfies the conditions C_0, C_1 and C_2 as defined in Section 3.1, is a trace automaton of A_G. In fact conditions C_1 and C_2 alone are already sufficient, since C_2, in combination with C_1, already implies C_0. Trace automata have been introduced in [God91]. They preserve safety properties, e.g. specified in form of assertions, of the complete state space. In [God96] they are defined as:

Definition 30 (Trace Automaton)

> Let A_G be the complete state space of a system. A reduced state space A_R for this system is a trace automaton for this system if, for all sequences w of transitions from the initial state s_0 in A_G, there exists a sequence w' of transitions from s_0 in A_R such that w' is a linearization of a trace defined by an extension of w, i.e. $w \in Pref([w']_{s_0})$.

The definition of trace automata directly coincides with the above main Theorem A.3.1. Consequently proving that theorem will also prove that A_R is a trace automaton of A_G, which means it will preserve all assertion violations of A_G.

A.3.1. Proof of Theorem A.3.1

First some auxiliary lemmata will be introduced. Finally the complete proof of the main theorem is presented.

Lemma A.3.2. *Let s be a state in A_R. Let $w = t_1..t_n$ be a trace in A_G leading to an error state from s. Let T be a persistent set in s. If $t_i \notin T$ for all $i \in \{1..n\}$, <u>then</u> $\forall a_i \in T : w \in Pref([a_i w]_s)$, <u>else</u> let t_k be the smallest $t_i \in T$, $w \in Pref([t_k w']_s)$ where $w' = t_1..t_{k-1}t_{k+1}..t_n$ is a trace in A_G.*

Proof. The proof will consider both cases, one where the *if* condition is true and one where it is false. Let $s = s_1 \xrightarrow{t_1} s_2 \xrightarrow{t_2} ...s_n \xrightarrow{t_n} s_{n+1}$ be the states that w goes through in A_G.

Case : $\forall i \in \{1..n\} : t_i \notin T$.

According to the Definition 10 of persistent sets, all transitions $a_i \in T$ are independent with t_j in s_j for all $j \in \{1..n\}$. Consequently $a_i w \in [wa_i]_s$ for all $a_i \in T$. Thus it immediately follows that $a_i w \in Pref([wa_i]_s)$ since $[wa_i]_s \subseteq Pref([wa_i]_s)$ by definition of *Pref*.

Case : $\exists i \in \{1..n\} : t_i \in T$.

Let t_k be the first $t_i \in T$ ($\forall t_i \in T : k \leq i$). Then by Definition 10 of persistent sets, t_k is independent with all t_j in s_j for $j \in \{1..k-1\}$. Consequently $t_k t_1..t_{k-1} \in [t_1..t_{k-1}t_k]_s$, which directly implies $t_k t_1..t_{k-1}t_{k+1}..t_n \in [t_1..t_{k-1}t_k t_{k+1}..t_n]_s$ since both traces are extended in the same way. Analogously to the above case it immediately follows that $t_k t_1..t_{k-1}t_{k+1}..t_n \in Pref([t_k w']_s)$ where $w' = t_1..t_{k-1}t_{k+1}..t_n$.

$\square$

Lemma A.3.3. *Let s be a state in A_R. Let $w = t_1..t_n$ be a trace in A_G leading to an error state from s. Let T be a persistent set in s. If $t_i \notin T$ for all $i \in \{1..n\}$, <u>then</u> $\forall t \in T : s \xrightarrow{tw}$ is also an error state in A_G, <u>else</u> let t_k be the smallest $t_i \in T$, let $w' = t_1..t_{k-1}t_{k+1}..t_n$ then $s \xrightarrow{t_k w'}$ is also an error state in A_G.*

Proof. The proof will consider both cases, one where the *if* condition is true and one where it is false.

Case : $\forall i \in \{1..n\} : t_i \notin T$.

According to Lemma A.3.2 $w \in Pref([a_i w]_s)$ for all $a_i \in T$. Since w leads to an error state from s in A_G, the trace $a_i w$ will also lead to an error state from s in A_G.

Case : $\exists i \in \{1..n\} : t_i \in T$.

According to Lemma A.3.2 $w \in Pref([t_k w']_s)$. Since w leads to an error state from s in A_G, the trace $t_k w'$ will also lead to an error state from s in A_G.

$\square$

Lemma A.3.4. *Let s be a state in A_R. Let $w = t_1..t_n$ be a trace in A_G leading to an error state from s. Let $w_a = a_1..a_m$ be a trace in A_R. Let $s = s_1 \xrightarrow{a_1} s_2 \xrightarrow{a_2} ...s_m \xrightarrow{a_m} s_{m+1} = s_a$ be the states it goes through in A_R. Let $t_i \notin r(s_j)$ and $r(s_j)$ be a persistent set in s_j for $i \in \{1..n\}$ and $j \in \{1..m\}$. Then w will lead to an error state from s_a in A_G and $w \in Pref([w_a \cdot w]_s)$.*

Proof. The proof proceeds by induction over the length of w_a.

IB : $|w_a| = 0$

In this case $s_a = s$. Thus w leads to an error state from s_a in A_G by assumption. And $w_a \cdot w = w$ thus clearly $w \in Pref([w]_s) = Pref([w_a \cdot w]_s)$.

IS : $|w_a| = x+1 = m$. Let $w_a = a_1..a_m$ as defined in the proof header.

According to Lemma A.3.3 w will lead to an error state in A_G from s_2. According to Lemma A.3.2 $w \in Pref([a_1 \cdot w]_{s=s_1})$. By assumption s_2 is in A_R and the rest $w'_a = a_2..a_m$ of w_a is a trace in A_R leading to s_a from s_2. And (also by assumption) $t_i \notin r(s_j)$ and $r(s_j)$ is a persistent set in s_j for $i \in \{1..n\}$ and $j \in \{2..m\}$. Thus by IH. w will lead to an error state from s_a in A_G and $w \in Pref([w'_a \cdot w]_{s_2})$. Together with the already established intermediate result $w \in Pref([a_1 \cdot w]_{s=s_1})$ it follows that $w \in Pref([w_a \cdot w]_{s=s_1})$.

$\square$

Lemma A.3.5. *Let s be a state in A_R. The reduction function r satisfies the conditions C_1 and C_2 as defined in Section 3.1.3. Let $w = t_1..t_n$ be a non-empty trace in A_G that leads to an error state from s.*

Then there exists a (possibly empty) trace $w_a = a_1..a_m$ that goes through the states $s = s_1 \xrightarrow{t_1} s_2 \xrightarrow{t_2} ...s_a \xrightarrow{t_a} s_{a+1} = s_a$ in A_R such that $w \in Pref([w_a \cdot w]_s)$ and w leads to an error state from s_a in A_G and at least one transition t_i of w is in $r(s_a)$.

Proof. Since by assumption the execution of w from s is defined in A_G, the first transition of w is enabled in s, so $t_1 \in en(s)$. Thus according to condition C_2 of the reduction function r, there exists a trace in A_R such that a state s' is reached from s with $t_1 \in r(s')$. A weaker condition is that there exists a trace in A_R from s such that a state s' is reached where any $t_i \in r(s')$ for $i \in \{1..n\}$. Let $w_a = a_1..a_m$ be such a (possibly empty) trace in A_R and $s = s_1 \xrightarrow{t_1} s_2 \xrightarrow{t_2} ...s_a \xrightarrow{t_a} s_{a+1} = s_a$ be the states it goes through in A_R. W.l.o.g. $\neg \exists t_i \in r(s_j)$ for $j \in \{1..m\}$ (else a prefix of w_a can be used). Condition C_1 ensures that all $r(s_j)$ are persistent sets for $j \in \{1..m\}$. According to the above Lemma A.3.4, w will also lead to an error state from s_a and $w \in Pref([w_a \cdot w]_s)$, which is the other result part of the proof. $\square$

Theorem A.3.6

Let s be a state in A_R. The reduction function r satisfies the conditions C_1 and C_2 as defined in Section 3.1.3. Let w be a trace in A_G that leads to an error state. Then there exists a trace w_r in A_R such that $w \in Pref([w_r]_s)$.

Proof. The proof proceeds by induction on the length of the trace w.

IB : $|w| = 0$.

The base case immediately holds with $w_r = w$.

IS : $|w| = x+1 = n$. Let $w = t_1..t_n$.

According to Lemma A.3.5 there exists a trace w_a in A_R such that a state s_a is reached in A_R from s, where w leads to an error state in A_G from s_a and $w \in Pref([w_a \cdot w]_s)$ and $\exists t_i \in r(s_a)$. Let t_k be the first such t_i.

According to Lemma A.3.2 there exists a trace $w' = t_1..t_{k-1}t_{k+1}..t_n$ in A_G such that $w \in Pref([t_k \cdot w']_{s_a})$. Since $w \in Pref([w_a \cdot w]_s)$ has already been established above, it follows that $w \in Pref([w_a \cdot t_k \cdot w']_s)$.

According to Lemma A.3.3 $t_k \cdot w'$ leads to an error state from s_a. Let s' be the successor of s_a when executing t_k. Since s_a is in A_R and $t_k \in r(s_a)$, s' is also a state in A_R. By definition of $w' = t_1..t_{k-1}t_{k+1}..t_n$, $|w'| = |w| - 1$. Thus by IH. there exists a trace w'_r in A_R such that $w' \in Pref([w'_r]_{s'})$.

Together with the already established intermediate result $w \in Pref([w_a \cdot t_k \cdot w']_s)$ it follows that $w \in Pref([w_a \cdot t_k \cdot w'_r]_s)$, where $w_a \cdot t_k \cdot w'_r$ is a path from s in A_R.

□

From the above theorem it immediately follows that Theorem A.3.1 is also valid, since the above theorem holds for every state in A_R, hence also for the initial state. Consequently A_R is a trace automaton of A_G and thus preserves all assertion violations.

A.4. SSR without POR Correctness Proof: Assertion Violation Preserving

This section provides a proof for the following theorem, that appeared in Section 5.2 as Theorem 5.2.1.

Theorem A.4.1

Let w be a trace in the complete state space A_G that leads to an error state s_e from the initial state s_0. Let A_R be a corresponding reduced state space using the reduction function from Definition 13. Then an error state s'_e will also be weakly reachable in A_R from s_0 by w such that $s_e \preccurlyeq s'_e$.

First some auxiliary results will be established. The following lemma states that once a state s is covered by another state s', executing a single transition t, which is enabled in s (and hence has to be enabled in s' due to $s \preccurlyeq s'$), preserves the covered relation in the resulting states.

Lemma A.4.2 (Result Coverage Preserving). *If* $s \preccurlyeq s'$ *then* $\forall t \in T$ *with* $t \in en(s)$ *it holds* $s \xrightarrow{t} \preccurlyeq s' \xrightarrow{t}$.

Proof. By assumption $s \preccurlyeq s'$, thus for all $w \in T^*$: $\perp (s \xrightarrow{w}) \implies \perp (s' \xrightarrow{w})$. Since w is an arbitrary trace, let w start with the transition t followed by an arbitrary trace w': $w = t \cdot w'$. So $\perp (s \xrightarrow{t \cdot w'}) \implies \perp (s' \xrightarrow{t \cdot w'})$, which is equivalent to $\perp ((s \xrightarrow{t}) \xrightarrow{w'}) \implies \perp ((s' \xrightarrow{t}) \xrightarrow{w'})$. Thus $s \xrightarrow{t} \preccurlyeq s' \xrightarrow{t}$. □

Theorem A.4.3

Let A_R be a reduced state space using the reduction function from Definition 13. Let s be a state in A_R and w a trace in A_G leading to an error state s_e in A_G. Then an error state s'_e will be weakly reachable from s in A_R by trace w, such that $s_e \preccurlyeq s'_e$ holds.

Proof. The proof proceeds by induction over the length of w.

IB $|w| = 0$

This case immediately holds, since the empty trace does not change the current state and the $\preccurlyeq$ relation is reflexive.

IS $|w| = x + 1 = n$.

Let $w = t_1..t_n = t_1 \cdot w'$, where t_1 is the first transition and w' the rest of w. Now according to Definition 13 of the reduction function for SSR there are two possible cases:

case $r(s) = \emptyset$

Since, by assumption, w is executable from s in A_G, $en(s) \neq \emptyset$. Thus according to Definition 13 there exists a state s' in A_R with $s \preccurlyeq s'$ and $r(s') = en(s')$. Since $s \preccurlyeq s'$ and w leads to an error state s_e from s in A_G, w will also lead to an error state s'_e from s' in A_G. Thus $t_1 \in r(s') = en(s')$. Let s_t be the successor of s' in A_R when executing transition t_1. The state s_t is weakly reachable from s, since $s \preccurlyeq s'$. The error state s'_e is reachable from s_t by w' in A_G. By IH. an error state s''_e is weakly reachable from s_t in A_R by trace w'. According to Lemma A.4.2 $s_e \preccurlyeq s'_e$ and $s'_e \preccurlyeq s''_e$ holds, which immediately implies $s_e \preccurlyeq s''_e$.

case $r(s) = en(s)$

Since w is executable in A_G from s, $t_1 \in en(s) = r(s)$. Let s_t be the successor of s in A_R when executing transition t_1. The state s_t is weakly reachable from s, since it is directly reachable from s. The error state s_e is reachable from s_t by w' in A_G. By IH. an error state s'_e is weakly reachable from s_t in A_R by trace w'. According to Lemma A.4.2 $s_e \preccurlyeq s'_e$ holds.

$\square$

The main theorem (Theorem A.4.1) of this section follows immediately from the above theorem, since it is valid for any state s in A_R, thus also for the initial state s_0.

A.5. SSR with POR Correctness Proof: Assertion Violations Preserving

This sections provides a proof for the following theorem introduced in Section 5.3 as Theorem 5.3.1.

Theorem A.5.1 (*Assertion violations preserving combined reduction function*)

> Let A_R be a reduced state space where the reduction function r satisfies the conditions C_1 and C_{2W} as defined in Section 5.3. Let w be a trace in A_G leading to an error state from the initial state s_0. Then there exists a trace w_r in A_R such that an error state is weakly reachable from s_0 in A_R.

The idea is to proof this theorem by induction over the length of the trace w. But first some auxiliary results are established.

Lemma A.5.2. *Let s be a state in A_R. Let $w = t_1..t_n$ be a trace in A_G leading to an error state from s. Let $w_a = a_1..a_m$ be a trace in A_R, such that a state s_a in A_R is weakly reachable from s by w_a. Let $s = s_1..s_{k+1} = s_a$ be the weakly reachable states that it goes through in A_R. Let $t_i \notin r(s_j)$ and $r(s_j)$ is a persistent set in s_j, for all $i \in \{1..n\}$ and for all $j \in \{1..k\}$. Then w will lead to an error state from s_a in A_G.*

Proof. The proof proceeds by induction over the length the trace w.

IB : $|w| = 0$.

In this case $s \preccurlyeq s_a$ by Definition 16 of weak reachability. Thus w leads to an error state from s_a in A_G by Definition 14 of $\preccurlyeq$.

IS : $|w| = x + 1 = m$.

In this case $s_1 \preccurlyeq s_2$ and $s_2 \xrightarrow{a_1}_R s_3$ and s_a is weakly reachable from s_3 by the rest $a_2..a_m$ of w_a by Definition 16 of weak reachability. Analogously to IB. w will lead to an error state from s_a in A_G. Let $s_2 = z_1 \xrightarrow{t_1} z_2 \xrightarrow{t_2} ...z_n \xrightarrow{t_n} z_{n+1}$ be the states w goes through in A_G. None t_i of w is in the persistent set $T = r(s_2)$ of s_2 by assumption. Thus by Definition 10 of persistent sets, a_1 is independent with t_j in z_j for $j \in \{1..n\}$. According to Definition 6 of trace equivalence $a_1 \cdot w \in [w \cdot a_1]_{s_2}$, which means executing a_1 before or after w from s_2 will lead to the same state. Consequently, w will lead to an error state from s_3 in A_G. Thus by IH. w will lead to an error state from s_a in A_G.

$\square$

Lemma A.5.3. *Let s be a state in A_R, which satisfies the conditions C_1 and C_{2W} as defined in Section 5.3. Let $w = t_1..t_n$ be a non-empty trace in A_G leading to an error state from s. Then there exists a (possibly empty) trace w_a in A_R such that a state s_a is weakly reachable from s in A_R, where at least one transition t_i of w is in $r(s_a)$ and w leads to an error state from s_a in A_G.*

Proof. By assumption w is non-empty and leads to an error state in A_G. Consequently, according to C_{2W}, there exists a trace $w_a = a_1..a_m$ such that a state s_a is weakly reachable from s by w_a in A_R, where at least one transition t_i of w is in $r(s_a)$, i.e. $t_i \in r(s_a)$. Let $s_1..s_{k+1} = s_a$ be the weakly reachable states that w_a goes through in A_R. Please notice that the number of states can be greater than the number of transitions (hence different indices are used), since the notion of weak reachability allows empty transition between states, i.e. if $s_x \preccurlyeq s_y$ then s_y is weakly reachable from s_x by the empty trace, however k is still finite. W.l.o.g. no transition t_i of w is in $r(s_j)$ (else one could use a prefix of the existing w_a) and condition C_1 ensures that $r(s_j)$ is a persistent set in s_j, for all $j \in \{1..k\}$. Consequently, according to the above Lemma A.5.2, w will also lead to an error state from s_a, which is the other result part of the proof. $\square$

These auxiliary lemmata together with some of those already introduced in Section A.3 can be used to proof the following main theorem.

Theorem A.5.4

Let s be a state in A_R. The reduction function r satisfies the conditions C_1 and C_{2W} as defined in Section 5.3. Let w be a trace in A_G leading to an error state from s. Then there exists a trace w_r in A_R such that an error state is weakly reachable from s in A_R.

Proof. The proof proceeds by induction on the length of w.

IB : $|w| = 0$

This case immediately holds with $w_r = w$.

IS : $|w| = x + 1 = n$. Let $w = t_1..t_n$.

According to Lemma A.5.3 there exists a trace w_a such that a state s_a is weakly reachable from s in A_R by w_a, and at least one transition t_i of w is in $r(s_a)$, and w will lead to an error state in A_G. Let t_k be the first such t_i. Let s' be the successor of s_a when executing t_k, s' is in A_R since s_a is in A_R and $t_k \in r(s_a)$. According to Lemma A.3.3 there exists a trace w' that will lead to an error state from s' in A_G and $|w'| = |w| - 1$. Thus by IH there exists a trace w'_r such that an error state is weakly reachable from s' in A_R. Consequently an error state is weakly reachable from s in A_R by the trace $w_a \cdot t_k \cdot w'_r$.

$\square$

From the above theorem it immediately follows that Theorem A.5.1 is also valid, since the above theorem holds for every state in A_R, hence also for the initial state.

A.6. Reduction Function Condition Proofs

This section provides proofs for the theorems 5.3.2 and the Lemma 3.1.2 introduced in Section 3.1.3 and Section 5.3. The proofs are available in Theorem A.6.4, Theorem A.6.2 and Lemma A.6.1, respectively.

Lemma A.6.1. *Condition C_2 implies condition C_0.*

Proof. To show $r(s) = \emptyset \iff en(s) = \emptyset$ both implication directions will be considered.

$\Rightarrow$ This immediately follows from condition C_2 by contraposition. Condition C_2 states that whenever a transition is enabled in s it will be explored from a state reachable from s, thus $r(s)$ cannot be empty: $en(s) \neq \emptyset \implies r(s) \neq \emptyset$. By contraposition $r(s) = \emptyset \implies en(s) = \emptyset$ follows.

$\Leftarrow$ The other implication direction holds by definition, because if no transition is enabled in s then obviously no transition can be explored in the reduced state space, which is a sub-space of the complete state space by definition.

$\square$

Theorem A.6.2

> The condition C_2 is implied by the condition C_1 and C_2^S.

Proof. Let s be a state in A_R. First there are two possible cases:

- $en(s) = \emptyset$: in this case C_2 is satisfied by definition.

- $en(s) \neq \emptyset$: this case is proved by contradiction. Let t be a transition that is enabled in s, i.e. $t \in en(s)$. According to C_2^S there exists a trace $w = t_1..t_n$ in A_R such that a fully expanded state s' is reachable, which means $r(s') = en(s')$. Let $s = s_1 \xrightarrow{t_1} s_2 \xrightarrow{t_2} ...s_n \xrightarrow{t_n} s_{n+1} = s'$ be the states w goes through in A_R. Now assume that $t \notin r(s_j)$ for $j \in \{1..n+1\}$.

  Since condition C_1 holds by assumption, all $r(s_j)$ are persistent sets. By Definition 10 of persistent sets, t_j is independent with t in s_j for $j \in \{1..n\}$. Consequently t must still be enabled in s', by Definition 4 of transition (in-)dependence. Since $s' = s_{n+1}$ is fully expanded $t \in r(s)$ should hold. This contradicts with the assumption $t \notin r(s_j)$ for $j \in \{1..n+1\}$. Consequently $t \in r(s_j)$ for some $j \in \{1..n+1\}$ which is exactly what condition C_2 states.

□

Lemma A.6.3. *Let s be a state in A_R, where the reduction function satisfies the conditions C_1 and C_{2W}^S. Let $w_a = a_1..a_m$ be a trace in A_R, such that a fully expanded state s_a is weakly reachable from s in A_R. Let $s = s_1..s_k = s_a$ be the weakly reachable states w_a goes through in A_R. Let $w = t_1..t_n$ be a non-empty trace leading to an error state from s in A_G. Then $t \in r(s_j)$ for some $j \in \{1..k\}$.*

Proof. The proof proceeds by induction over the length of w.

IB $|w| = 0$

> In this case $s \preccurlyeq s_a$ by Definition 16 of weak reachability. Since w leads to an error state from s in A_G, w will also lead to an error state from s_a in A_G, by Definition 14 of $\preccurlyeq$. Thus the first transitions t_1 of w is enabled in s_a. By assumption s_a is fully expanded, i.e. $r(s_a) = en(s_a)$, thus $t_1 \in r(s_a)$.

IS $|w| = x + 1 = m$

> In this case $s_1 \preccurlyeq s_2$ and $s_2 \xrightarrow{a_1}_R s_3$ and s_a is weakly reachable from s_3 by the rest $a_2..a_m$ of w_a by Definition 16 of weak reachability.
>
> Analogously to the IB. case, w will lead an error state from s_2 in A_G. Let $s_2 = z_1 \xrightarrow{t_1} z_2 \xrightarrow{t_2} ...z_1 \xrightarrow{t_1} z_{1+1}n$ be the states w goes through in A_G. Now there are two possible cases:
>
> – $\exists i \in \{1..n\} : t_i \in r(s_2)$: in this case the proof is complete.
>
> – $\forall i \in \{1..n\} : t_i \notin r(s_2)$: the condition C_1 ensures that $r(s_2)$ is a persistent set in s_2. None t_i of w is in the persistent set $r(s_2)$ of s_2 in this case. Thus by Definition 10 of persistent sets, a_1 is independent with t_j in z_j for $j \in \{1..n\}$. According to Definition 6 of trace equivalence $a_1 \cdot w \in [w \cdot a_1]_{s_2}$, which means executing a_1 before or after w from s_2 will lead to the same state. Consequently, w will lead to an error state from s_3 in A_G. Thus by IH. $t \in r(s_j)$ for some $j \in \{3..k\}$.

□

Theorem A.6.4

> The condition C_{2W} is implied by the conditions C_1 and C_{2W}^S.

Proof. Let s be a state in A_R, where the reduction function satisfies the conditions C_1 and C_{2W}^S. Let w be a non-empty trace leading to an error state from s in A_G. Let t be the first transition of w. Since w is executable from s in A_G, t is enabled in s, i.e. $t \in en(s)$. According to C_{2W}^S, there exists a trace w_a in A_R, such that a fully expanded state s_a is weakly reachable from s in A_R. Then according to Lemma A.6.3 there exists a *weakly reachable* state s' from s in A_R such that $t \in r(s')$.   □

A.7. Refined Assertion Violation Preserving Exploration

This section presents a more detailed version of the AVPE in Algorithm 8 presented in Section 3.2.3. It is a non-recursive implementation that explicitly handles multiple successor states of a single transition due to symbolic branch conditions and the different simulation phases of the SystemC runtime as presented in Section 2.1. The current simulation phase can be retrieved from a state s by accessing $s.simState$. The predicates $s.simState.isEval$, $s.simState.isNotify$ and $s.simState.isEnd$ will return True, if the simulation in state s is in the *evaluation* phase, *notification* phase or has completed. The *notification* phase, by convention in this thesis, comprises the *update*, *delta notify* and optional *timed notification* phases, as defined in Section 2.1.1.

The extended algorithm is shown in Algorithm 22. The algorithm performs a DFS search. It manages a global (search) stack and a set H of already visited states. Both are initially empty. Every state is associated with additional informations used by the search algorithm. The sets *working* and *done* contain transitions that need to be explored and have already been explored respectively. The flag *expanded* to decide whether the state has already been examined. Due to the non recursive implementation, states can be re-pushed multiple times on the stack, if they have multiple outgoing transitions. However, some actions need only be performed when the state is pushed for the first time, e.g. checking if the state has already been visited. The flags *unfinished* and *safe* are used to incorporate the cycle proviso as described in Section 3.2.2. By default all flags are set to False and the transition sets are both empty. The *initialize* function is used to associate these default informations with each state.

main loop The algorithm begins by initializing the initial state and then pushes it on the search stack. Now it will loop until the search stack is not empty. In every iteration the top state s of the search stack is removed. First it will be checked whether s has already been expanded. This is the case if s was already on the stack and has been re-pushed. In this case it will be checked whether the working set of s is empty, which means the exploration of s is complete. If it is, s will be backtracked, else a transition from s will be executed by calling the *runStep* function. Else s was not yet expanded in Line 7, thus it will first be expanded and then a transition from s will be executed. In the following these three main functions, namely *expand*, *backtrack* and *runStep* will be further explained one after another. Each of them gets the currently examined state s as argument.

expand The *expand(s)* function first checks whether s has already been visited. If so, the already visited state v is retrieved. If v is safe, then all states that can reach it are also marked safe. Else v is marked as unfinished, thus it will eventually be refined during backtracking. An exception is used to stop the execution of this path, since s has already been visited, and transfer the control back to the main loop, which will then examine the next state on the stack. Only states where the kernel is in the *evaluation* or notification phase of the simulation are stored in the cache. This is just a design decision, since these phases are the important one for the simulation, as described in Section 2.1.1. Else s has not yet been visited. First s is added to the set of visited states. Then if s is in the *evaluation* phase of the simulation, a persistent set will be computed in s, starting from some enabled transition, and assigned to $s.working$. If s is fully expanded or in the *notification* phase of the simulation, in which case it is fully expanded by definition, s and all states that can reach it will be marked safe.

backtrack The *backtrack(s)* function will check whether some relevant transition may have been ignored from the state s. This is the case if s lies on a cycle of states (s.unfinished is true) and does not reach a fully expanded state (s.safe is not true). In this case s will be refined. The implementation shown in Algorithm 22 will simply fully expand the state s and than mark s and all states on the search stack, since they can reach s, as safe. States are marked in reverse order as an optimization. This allows to stop the marking once a safe state is detected, since all states below on the stack must already be safe.

Algorithm 22: Refined AVPE

```
1  H ← Set()
2  states ← stack()

3  pushState(initialState)
4  while |states| > 0 do
5  │  s ← states.pop()
6  │  try
7  │  │  if s.expanded then
8  │  │  │  if |s.working| = 0 then
9  │  │  │  │  backtrack(s)
10 │  │  │  else
11 │  │  │  │  runStep(s)
12 │  │  else
13 │  │  │  expand(s)
14 │  │  │  runStep(s)
15 │  catch AbortPath
16 │  │  pass

17 procedure initialize(s) is
18 │  s.expanded ← False
19 │  s.safe ← False
20 │  s.unfinished ← False
21 │  s.working ← {}
22 │  s.done ← {}

23 procedure expand(s) is
24 │  s.expanded ← True
25 │  if s.simState.isEval ∨ s.simState.isNotify then
26 │  │  if s ∈ H then
27 │  │  │  v ← H[s]
28 │  │  │  if v.safe then
29 │  │  │  │  markAllSafe(s)
30 │  │  │  else
31 │  │  │  │  v.unfinished ← True
32 │  │  │  raise AbortPath
33 │  │  else
34 │  │  │  H.add(s)
35 │  │  if s.simState.isEval then
36 │  │  │  assert (|s.enabled| > 0)
37 │  │  │  t ← selectElement(s.enabled)
38 │  │  │  s.working ← persistentSet(s, t)
39 │  │  │  if |s.working| = |s.enabled| then
40 │  │  │  │  markAllSafe(s)
41 │  │  else
42 │  │  │  markAllSafe(s)
```

```
43 procedure markAllSafe(head) is
44 │  head.safe ← True
45 │  for s ∈ reversed(states) do
46 │  │  if s.safe then
47 │  │  │  break
48 │  │  s.safe ← True

49 procedure pushState(s) is
50 │  initialize(s)
51 │  states.push(s)

52 @require: s.expanded
53 procedure repushState(s) is
54 │  states.push(s)

55 function selectNextTransition(s) is
56 │  t ← selectElement(s.working)
57 │  s.working.remove(t)
58 │  s.done.add(t)
59 │  return t

60 procedure runStep(s) is
61 │  if s.simState.isEnd then
62 │  │  raise AbortPath
63 │  else if s.simState.isEval then
64 │  │  n ← s.cloneState()
65 │  │  repushState(s)
66 │  │  t ← selectNextTransition(s)
67 │  │  n.executeTransition(t)
68 │  else if s.simState.isNotify then
69 │  │  n ← s.cloneState()
70 │  │  n.executeNotificationPhase()
71 │  else
72 │  │  s.continueExecution()
73 │  │  n ← s
74 │  if n.simState.isFork then
75 │  │  pushState(n.forked)
76 │  pushState(n)

77 procedure refineWorkingSet(s) is
78 │  s.working ← s.enabled \ s.done
79 │  markAllSafe(s)

80 procedure backtrack(s) is
81 │  if s.unfinished ∧ ¬s.safe then
82 │  │  refineWorkingSet(s)
83 │  │  if |s.working| ≠ ∅ then
84 │  │  │  repushState(s)
```

Remark. A different refinement implementation would be to compute a new persistent set T starting with a so far unexplored transition $t \in enabled(s)$ *s.done*. The working set would be set to T, i.e. *s.working* $= T$. The *markAllSafe* call would then only be issued, if *s.done* $\cup$ *s.working* equals *enabled*(*s*).

runStep The *runStep(s)* function starts or continues the execution of a single transition. The actual behaviour depends on the current simulation phase and whether the execution of a transition has been interrupted. A transition will be interrupted, whenever a branch with a symbolic condition c is executed where both c and $\neg c$ are satisfiable with the current path condition. The execution of the current transition will *fork* into two independent execution paths s_T and s_F. The path condition of s_T one state with c and the other one with $\neg c$ as described in Section 2.3. These states are said to be *forked*. The predicate *s.simState.isFork* will return *True* for such states. Given a state s, the forked state can be accessed via *s.forked*.

Now the behaviour of the *runStep* function can be described as follows: If the simulation is completed, the function will simply return. Else if the simulation in state s is in the *evaluation* phase, s will be cloned into the state n. Cloning s ensures that the stored copy in the visited set is not modified and other transitions can be explored from s later on. The original state s is re-pushed on the stack, thus it will be examined from within the main loop again. At that point s will be either backtracked or a different transition will be explored from s. Next, an unexplored transition from the persistent set in s is selected and explored from the state n. Selection happens from s, to modify the search informations associated with s and execution happens from n to keep the actual state s unmodified. Similarly the state s will be cloned if the simulation is in the *notification* phase. The notification phase, which comprises the *update, delta notify* and optional *timed notification* phases, is executed from the cloned state n. All other simulation phases, or an interrupted transition, have only a single choice, thus they will simply continue execution. Since, by a design decision, these states are not stored in the visited set, they are not cloned, but directly transfered to a new state. In all cases the successor state n is reached, which is pushed on the search stack. If the transition execution had been interrupted due to a symbolic branch, the forked state will also be pushed on the stack for further examination. Thus the search algorithm just puts both of them onto the search stack for further exploration. Doing so seems to be a natural choice since both forked states, which represent execution paths, are completely independent of one another. Since all of its predecessor states are below them on the stack, such an implementation is compatible with the cycle proviso C_2.

A.8. SDPOR Algorithm

This section presents the complete SDPOR algorithm, that will fall back to static persistent sets, whenever the computation of a non trivial persistent set fails dynamically. In this case the implementation of a cycle proviso becomes mandatory to preserve properties more elaborate than deadlocks. The idea on how to integrate static persistent sets, has already been briefly discussed as the stateless DPOR algorithm has been presented (Section 33) and some additional notes have been given for the C-SDPOR extension (Section 4.3). For clarity this section does present the complete combined algorithm for the final stateful DPOR version SDPOR.

Assuming that an algorithm is available that can compute static persistent sets (e.g. one of those presented in Section 2.6.2), the static persistent sets can be integrated quite intuitively. When a state s is visited the first time, an arbitrary enabled transition t is selected to initialize the working set *working* of s. Additionally s is associated with a persistent set around t (Line

Algorithm 23: SDPOR Algorithm combined with Static Persistent Sets

Input: Initial state

1 $H \leftarrow Set()$
2 $G_v \leftarrow VisibleEffectsGraph()$
3 $P \leftarrow Trace(initialState)$
4 $explore()$

5 **procedure** *explore()* **is**
6   s ← last(P)
7   **if** $s \in H$ **then**
8        $v \leftarrow H[s]$
9   **if** *v.safe* **then**
10   *P.markAllSafe()*
11   **else**
12   v.unfinished ← True
13   **if** $v \in P.statesOfLastDeltaCycle()$ **then**
14   c ← collectStateCycle(v)
15   expandStateCycle(c)
16   **else**
17   *extendedBacktrackAnalysis(v)*
18   **else**
19       $H.add(s)$
20   s.done ← {}
21   **if** $enabled(s) \neq \emptyset$ **then**
22          $t \leftarrow selectAny(enabled(s))$
23   s.persistentSet ← staticPersistentSet(s, t)
24   s.working ← {t}
25   **while** $|s.working| > 0$ **do**
26   *exploreTransitions*(s, s.working)
27   **if** $s.done = enabled(s)$ **then**
28   *P.markAllSafe()*
29   **else if** $\neg s.safe \land s.unfinished$ **then**
30   *P.markAllSafe()*
31   *exploreTransitions*(s, $enabled(s) \setminus s.done$)
32       $H.backtrack(s)$

33 **procedure** *expandStateCycle(c)* **is**
34   **for** $s \in c$ **do**
35   *addBacktrackPoints(s, s.persistentSet)*

36 **procedure** *collectStateCycle(v)* **is**
37   c ← {}
38   **for** $s \in P.statesInReversedOrder()$ **do**
39       $c \leftarrow c \cup s$
40   **if** $s = v$ **then**
41   **break**
42   **return** c

43 **procedure** *addBacktrackPoints(s, T)* **is**
44   *s.working* ← $T \setminus s.done$

45 **procedure** *exploreTransitions(T)* **is**
46   **for** $t \in T$ **do**
47   s.working ← $s.working \setminus t$
48   s.done ← $s.done \cup t$
49       $n \leftarrow succ(s, t)$
50       $e \leftarrow visibleEffects(s, t)$
51       $G_v.addEdge(id(s), id(n), e)$
52       $P.push(e, n)$
53       $explore()$
54       $P.pop()$
55   *normalBacktrackAnalysis*(s, e)

56 **procedure** *normalBacktrackAnalysis(s, e)* **is**
57   *addBacktrackPoints*(s, e.disables)
58   *commonBacktrackAnalysis*(e)

59 **procedure** *extendedBacktrackAnalysis(v)* **is**
60   **for** $e \in G_v.reachableTransitionEffects(id(v))$ **do**
61   *commonBacktrackAnalysis*(e)

62 **procedure** *commonBacktrackAnalysis(e)* **is**
63   **for** $s_p, e_p \in transitionsLastDeltaCycle(P)$ **do**
64   **if** $areDependent(e_p, e)$ **then**
65          $E \leftarrow \{thread(t) \mid t \in s_p.persistentSet\}$
66   **if** $thread(e) \in E$ **then**
67             $T \leftarrow \{next(s_p, thread(e))\}$
68   **else**
69             $T \leftarrow s_p.persistentSet$
70   *addBacktrackPoints*(s_p, T)

23).

Dependent transitions are only added to the *working* set of s during backtracking, if they are in the persistent set initially associated with s (Line 67)[4]. If the analysis fails to infer a non trivial persistent set in a state (Line 69), or when a cycle of states is detected and expanded Line 35, the states are not fully expanded, but restricted to the initially computed persistent set. Only in Line 31 all enabled transitions will be explored that have not yet been explored. This is required to solve the ignoring problem as has already been discussed in Section 3.2.2. Thus it cannot be restricted to the initially computed persistent set.

A.9. Solving the Ignoring Problem implicitly

In order to preserve properties more elaborate than deadlocks in partial order reduced exploration, the ignoring problem has to be solved. It turns out that under certain conditions the DPE and A-SDPOR algorithms, which do not explicitly solve the ignoring problem, already implicitly solve it. One condition is that a simulation semantics is used, that is similar to those of IVL (which corresponds to the semantics of SystemC) programs. The other condition is, that the exploration algorithms compute persistent sets in a *compatible* way. The static conflicting transition (CT) algorithm and the backtracking analysis of the A-SDPOR algorithm, when it does not fall back to persistent sets, are compatible.

The following section shows that the DPE+CT algorithm solves the ignoring problem. The next section shows it for the A-SDPOR algorithm. Finally the usefulness of these results is discussed.

A.9.1. Static Partial Order Reduction

First some auxiliary lemmata will be introduced, then it will be shown that the DPE+CT, also preserves assertion violations for IVL programs.

Lemma A.9.1. *For every cycle of states that is explored by the DPE+CT algorithm in the same delta cycle, there exists at least a state s in the cycle of states such that s is fully expanded.*

Proof. Transitions are not interrupted preemptively. They are always executed until the next context switch or the end of their thread. Consequently whenever a transition of thread p is executed from state s reaching state s', the next transition of p is disabled (or not available if the end has been reached) in s'.

The only way to re-enable a thread p in the same delta cycle is to use an immediate notification (according to the semantics of the IVL). This requires that (a) p waits for an event and (b) another thread immediately notifies that event after p has been executed (else the notification would be lost).

These two requirements need to be satisfied in order to form a cycle of states in the same delta cycle. Let s be a state in delta cycle d. Let $w = t_1..t_n$ be a trace that goes through the cycle of states $s = s_1 \xrightarrow{t_1} s_2 \xrightarrow{t_2} ...s_n \xrightarrow{t_n} s_{n+1} = s$. In order two form a cycle of states, at least one transition has to be executed, thus w is not empty. The cycle requirements (a) and (b) immediately imply that:

- At least two threads are required to form a cycle of states in the same delta cycle.

---

[4]Actually if their thread corresponds to a thread of any transition in the persistent set, as defined in [FG05].

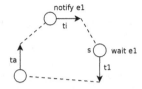

Figure A.1.: Cycle of states in same delta cycle

- Every transition in the cycle executes a wait event statement as context switch.

- There exists a transition t_a, which might be equal to t_1, in w such that $next(s_{a+1}, thread(t_a))$ $= t_1$ and $thread(t_a) = thread(t_1)$.

- There exists a transition t_i in w after t_a, such that $thread(t_i) \neq thread(t_a)$ and t_i immediately notifies the event that t_1 waits for.

So t_1 and t_a are transitions that belong to the same thread (they can also be equal). Without loss of generality t_a is the last transition of the $thread(t_a) = thread(t_1)$ that is executed before t_1. Thus all transitions $t_{a+1}..t_n$ are from a different thread. Since t_i is executed after t_a, the $thread(t_a)$ is not enabled in s_i. Since t_a waits for an event that t_i immediately notifies, t_a is dependent with t_i in s_i. The principle is shown in Figure A.1.

Consequently every valid (statically precomputed) dependency relation will capture this dependency. The *conflicting transitions* (CT) algorithm will return the complete set of enabled transitions whenever a disabled transition is introduced in the persistent set. Since t_i is in the persistent set of s_i and t_a is disabled in s_i, the CT algorithm must have returned all transitions enabled in s_i. Thus s_i, which is a state of the cycle, is fully expanded.   □

Lemma A.9.2. *For every cycle of states explored by the DPE+CT algorithm there exists a state s in the cycle of states, such that s is fully expanded.*

Proof. According to Lemma A.9.1 for every cycle of states in the same delta cycle, at least one state is fully expanded. A cycle of states that spans multiple delta cycles, must by definition of delta cycles contain a fully expanded state. The reason is that between every two delta cycles there exist at state s, such that only the designated transition t_N is enabled in s. Basically the transition t_N denotes the execution of a notification phase from s. The successor state will then be in the *eval* phase of the next delta cycle. So in both cases, the cycle of states spans multiple delta cycles or is contained in a single delta cycle, at least one state of the cycle is fully expanded.   □

Theorem A.9.3

> The deadlock preserving exploration (DPE) algorithm, presented in Section 3.2.1, where the conflicting transitions (CT) algorithm is used to compute static persistent sets, denoted as DPE+CT, will also preserve all assertion violations for any IVL program.

Proof. It has already been show in Section 3.2.1 that the DPE algorithm satisfies condition C_0 and C_1. If the state space does not contain any cycles, the DPE algorithm will for every state s eventually reach a terminal state, since the state space is finite. By definition a terminal state is fully expanded. According to Lemma A.9.2 in every cycle of states there exists a state, which is fully expanded. Consequently every state that can reach any state of the cycle can reach that fully expanded state. It follows immediately that from every state s the DPE algorithm will reach a state s' which is fully expanded. So condition C_2^S is satisfied too, which implies the weaker condition C_2. Thus all assertion violations are preserved. □

A.9.2. Dynamic Partial Order Reduction

This section shows that the standard A-SDPOR algorithm already satisfies the cycle proviso C_2^S. It will be shown that for every cycle of states at least one state will be fully expanded. First for cycles of states limited to a single delta cycle and then for cycles of states spanning possibly multiple delta cycles. The A-SDPOR algorithm is only sound for acyclic state spaces. But the C-SDPOR algorithm builds on top of the A-SDPOR algorithm. Currently the C-SDPOR algorithm fully expands all (relevant) states in each cycle, thus clearly solving the ignoring problem. But once a more sophisticated implementation is used, the ignoring problem needs to be considered again. With the results obtained in this section, the cycle proviso would still be preserved[5].

Lemma A.9.4. *For every cycle of states in the same delta cycle explored by the A-SDPOR algorithm, there exists a state s in the cycle of states, such that s is fully expanded, that means $s.done = enabled(s)$ when s is finally backtracked.*

Proof. In the proof of Lemma A.9.2 it has been shown that for every cycle of states that appears in the same delta cycle there always exists two transitions t_a and t_i, such that t_a is dependent with t_i in s_i and $thread(t_a)$ is not enabled in s_i (so neither transition of $thread(t_a)$ is enabled in s_i). Whenever a cycle of states is detected, the backtrack analysis of the A-SDPOR algorithm will, by definition of the A-SDPOR algorithm, compare all transitions of the cycle with each other. Thus t_a is compared with t_i in state s_i. So the dependency is detected. Since $thread(t_a)$ is not enabled in s_i, which means that neither transition of $thread(t_a)$ is enabled, the next transition of $next(s_i, thread(t_a))$ is also not enabled. Hence the algorithm will add all enabled transitions to the *working* set of s_i which have not already been explored. Thus all enabled transitions will be explored from s_i. □

Lemma A.9.5. *For every cycle of states explored by the A-SDPOR algorithm there exists a state s in the cycle of states, such that s is fully expanded, that means $s.done = en(()s)$ when s is finally backtracked.*

Proof. According to Lemma A.9.4 for every cycle of states in the same delta cycle, at least one state is fully expanded. A cycle of states that spans multiple delta cycles, must by definition of delta cycles contain a fully expanded state, as described in Lemma A.9.2. □

Thus the A-SDPOR algorithm would fully expand a state in each cycle of states, therefore satisfying the cycle proviso C_2^S. Since the C-SDPOR algorithm is built on top of the A-SDPOR algorithm, the same result applies to it. Thus whenever a more sophisticated algorithm is employed in the C-SDPOR algorithm to handle cycles of states, the cycle proviso will still be preserved.

---

[5]The proviso could be violated when a more sophisticated backtracking dependency analysis is employed, but that part of the algorithm is not directly dependent with the way cycles of states are handled to ensure that persistent sets are explored in each state.

A.9.3. Discussion

First of all these algorithms discussed in the previous sections do not solve the ignoring problem in general, but only when certain conditions are met. One of them is, that a simulation semantics is used that is similar to those of IVL (which corresponds to the semantics of SystemC) programs. The other condition is, that the exploration algorithms compute persistent sets in a *compatible* way. It turns out that the static CT algorithm and the backtracking analysis of the A-SDPOR algorithm, when it does not fall back to persistent sets, are compatible.

So it might be unnecessary to use an extended algorithm that explicitly solves the ignoring problem. This offers some slight advantages. First of all a less complex algorithm has to be implemented or if a stateless algorithm is already available it can be immediately be extended to a stateful version, simply by tracking already visited states. And second the additional performance and memory overhead[6] is removed, since it is not necessary to keep track of *safe* and *unfinished* states and traverse the search stack to mark *safe* states.

Though it seems that using the DPE+CT algorithm will not yield any further reduction compared to the AVPE+CT algorithm. The reason is that the AVPE algorithm is based on the DPE algorithm. So using AVPE+CT there also will be a state s for every cycle of states, such that all enabled transitions in s are explored. Thus the specific extensions to solve the ignoring problem explicitly will not introduce any further refinements, since every state of the cycle has already to be marked *safe*.

So altogether the unextended algorithm that solves the ignoring problem implicitly offers only negligible lower resource usage and does not offer a better reduction than a more sophisticated algorithm that requires an explicit extension to solve the ignoring problem. The AVPE+STUB combination normally performs a better state space reduction than the DPE+CT combination. Though the implementation of DPE+CT is less complex and only requires a valid dependency relation between transitions, a can-enable relation is not required.

A.10. Correctness Proof of the Exact Symbolic Subsumption

Different algorithms have been presented to detect covered/subsumed states. Basically they can be grouped into solver-based and explicit methods. The ESS method is the most general of them. All other algorithms can be reduced upon it, as described in Section 6.6. Thus it is sufficient to prove the ESS method correct.

The idea is to show that whenever the ESS method detects that a state s_1 is covered by a state s_2 then the property $s_1 \sqsubseteq s_2$, that will be defined in the following, holds too. And next that $s_1 \sqsubseteq s_2$ implies $s_1 \preccurlyeq s_2$. The definition of $s_1 \preccurlyeq s_2$ appeared in Definition 14 and is repeated here for convenience, the definition of $s_1 \sqsubseteq s_2$ will be shown right after it.

Definition 31 (*Result Coverage*)

> A state s_1 is (result) covered/subsumed by a state s_2, denoted as $s_1 \preccurlyeq s_2$, iff for every *trace* w that leads to an error state from s_1, w also leads to an error state from s_2.

---

[6]Though the slight additional memory requirement is meanwhile negligible.

Definition 32

> A state s_1 is covered by a state s_2, denoted as $s_1 \sqsubseteq s_2$, iff $((C_1 \text{ and } C_2) \text{ or } C_3)$. With:
> $C_1 := s_1 \sim s_2$
> $C_2 := ESS_{\preccurlyeq}(s_1, s_2)$
> $C_3 := s_1$ is a terminal state (e.g. $pc(s_1)$ is unsatisfiable or no more transition is enabled)

In the following the relation $\sqsubseteq$ is referred to, when it is said that a state covers another one. The relation $\preccurlyeq$ will always be explicitly specified. Since the ESS algorithm is used for matching, the above condition C_2 holds by definition. Furthermore condition C_1 is also satisfied, since the ESS algorithm (symbolic state comparison in general) will only be applied if two states are structurally compatible. Condition C_3 is useful for the following proof, because if no transition is executed from a state s, then $s \preccurlyeq s'$ for all states s' by definition of $\preccurlyeq$.

Now it is left to show that $s_1 \sqsubseteq s_2$ implies $s_1 \preccurlyeq s_2$. The proof is outlined in the following. The idea is to show that for every trace $w \in T*$, either its execution from s_2 will lead to an error state, or its execution from s_1 will not lead to an error state and the property $s_1 \sqsubseteq s_2$ is preserved in the successor states $s_1 \xrightarrow{w} \sqsubseteq s_2 \xrightarrow{w}$. This property will be called (trace) error coverage. First it will be shown that executing a single statement preserves the above properties (single step error coverage). This will then be extended to multiple statements (multi step error coverage). Its direct implication is that whenever the execution of a statement reaches an error state in s_1 it will also reach an error state in s_2.

These proofs rely on three helper lemmas. The first one states that extending the path condition with an expression constructible in both states preserves the covered relation. The second one that assigning a value to a variable preserves the covered relation. And the third, that any condition that can be constructed in either state and is satisfiable in s_1 is also satisfiable in s_2. All of them follow quite naturally from the definition of the ESS method, since every value that can be constructed in s_1 can also be constructed in s_2.

The next section defines and proves the helper theorems. The section thereafter defines and proves single step-, multi step- and trace error coverage. This directly implies the desired property $s_1 \sqsubseteq s_2 \implies s_1 \preccurlyeq s_2$.

A.10.1. Preliminary Part

This section provides and proves some basic theorems which are used to prove the above main theorem. First some common definitions will be introduced.

Definition 33 (*Linked Expression*)

> An expression μ is linked between two states s_1 and s_2, if it can be constructed on both states using the same operations. The expression constructed on state s_1 is denoted as $s_1[\mu]$ and the one constructed on s_2 is denoted $s_2[\mu]$.

A sufficient condition to be able to construct linked expressions is that both states are structurally compatible $s_1 \sim s_2$ or a stronger condition is that one state is covered by the other $s_1 \sqsubseteq s_2$.

Definition 34 (*Value Covered*)

> Let $s_1 \sqsubseteq s_2$ and μ be an arbitrary linked expression in s_1 and s_2. If $s_1[\mu]$ can evaluate to a concrete value v (denoted as $v \in s_1[\mu]$) and $s_1 \sqsubseteq s_2$ then $s_2[\mu]$ can also evaluate to v. It will be denoted as $s_1[\mu] \subseteq s_2[\mu]$.

Definition 35 (*Expression Covered*)

> Let $s_1 \sqsubseteq s_2$ and μ be an arbitrary linked expression in s_1 and s_2. Then $s_1 \sqsubseteq (\mu)s_2$ states that $s_1[\mu] \subseteq s_2[\mu]$ while still preserving $ESS_{\preccurlyeq}(s_1, s_2)$ for every value of $s_1[\mu]$.

Its like adding a new global variable slot (with the same unique name) to s_1 and s_2 and assigning $s_1[\mu]$, $s_2[\mu]$ to it resulting in the states $\tilde{s}_1$ and $\tilde{s}_2$. Then $s_1 \sqsubseteq (\mu)s_2$ iff $\tilde{s}_1 \sqsubseteq \tilde{s}_2$.

Definition 36

> A condition c is satisfiable in a state s denoted as $sat(s, c)$ iff $pc(s) \wedge c$ is satisfiable.

Definition 37 (*Sat Covered*)

> State s_1 is satisfiability covered by state s_2, denoted as $sat \sqsubseteq (s_1, s_2)$, iff for every linked condition c (between s_1 and s_2) it holds that $sat(s_1, s_1[c]) \implies sat(s_2, s_2[c])$.

Lemma A.10.1 (Expression Preserving). *Let μ be an arbitrary expression that can be constructed in state s_1, denoted as $s_1[\mu]$. If $s_1 \sqsubseteq s_2$ then $s_1 \sqsubseteq (\mu)s_2$.*

Proof. Let μ be a value that is constructed using the same operations in both states s_1 and s_2, denoted as $s_1[\mu]$ and $s_2[\mu]$ respectively. This is possible because s_1 and s_2 are structurally compatible ($s_1 \sim s_2$), by def. of $s_1 \sqsubseteq s_2$.

By structural induction over μ it will be shown that $s_1 \sqsubseteq (\mu)s_2$ is valid. The base cases for the induction are the literal expression constructors. The operators will be handled in the induction step. The expressions considered in the proof are shown in Figure A.2. They represent the relevant expressions in the SystemC IVL.

IB: case $\mu = x$, where x is a concrete value ($x = int|bool|char$).

> This case is satisfied because $s_1[x] = s_2[x]$ is always true without further assumptions.

case $\mu = id$

> In this case, the identifier will be resolved and the corresponding variable value v will be returned. Since the same identifier is used in both states at the same program location, the same variable slot (which contains the variable value) will be selected. By assumption $s_1 \sqsubseteq s_2$. Thus $vars(s_1)[id] \subseteq vars(s_2)[id]$ for any identifier id.

case $\mu = $?(type)

> A fresh (fully unconstrained) symbolic literal with the same type can evaluate to the same values in both states. Since μ is fresh it will not interfere with any existing value.

case $\mu = $ new <type>

> A newly allocated pointer will have the same default value on both states. Which is either the concrete value *undefined*, the numeric literal zero, or a fresh symbolic literal. Either case has already been handled.

IS: case $\mu = a + b$

> By IH. $s_1 \sqsubseteq (a)s_2$ and $s_1 \sqsubseteq (b)s_2$. Thus it immediately follows that $s_1 \sqsubseteq (\mu = a + b)s_2$.
>
> This proof case is applicable to any n-ary expression, not just binary addition. Thus all other binary- and unary- expressions as well as the array index access expression are satisfied too.

⟨*expr*⟩ ::= ⟨*compound-expr*⟩ | ⟨*literal-expr*⟩

⟨*literal-expr*⟩ ::= int | bool | char
　　　　 | identifier
　　　　 | symbolic-literal
　　　　 | new ⟨*type*⟩

⟨*compound-expr*⟩ ::= ⟨*binary-expr*⟩
　　　　 | ⟨*unary-expr*⟩
　　　　 | object=⟨*expr*⟩ [index=⟨*expr*⟩]
　　　　 | object=⟨*expr*⟩ . member=id

　　　　 | new ⟨*type*⟩ [length=⟨*expr*⟩]

⟨*unary-expr*⟩ ::= - ⟨*expr*⟩
　　　　 | * ⟨*expr*⟩
　　　　 | ! ⟨*expr*⟩

⟨*binary-expr*⟩ ::= ⟨*expr*⟩ (+ | - | * | /) ⟨*expr*⟩
　　　　 | ⟨*expr*⟩ (&& | ||) ⟨*expr*⟩
　　　　 | ⟨*expr*⟩ (< | ==) ⟨*expr*⟩

Figure A.2.: Excerpt of relevant literal and compound expression constructors in the IVL

case μ = new a=<type> [b=<expr>]

The same type is used on both states. Similarly to the pointer allocation operator, new array elements are either initialized to the concrete value *undefined*, the numeric literal zero, or introduce (one or more) fresh symbolic literals. Thus every element of the array can evaluate to the same value in both states. By IH. $s_1 \sqsubseteq (b)s_2$. Thus the array cannot be smaller in state s_2 than in s_1. Consequently $s_1 \sqsubseteq (\mu)s_2$ is satisfied.

case μ = a=<expr> . b=<id>

The same identifier is used on both states, since the same expression is evaluated. By IH. $s_1 \sqsubseteq (a)s_2$, thus if a can evaluate to a specific value in s_1 it can evaluate to the same value in s_2. Consequently $s_1 \sqsubseteq (\mu)s_2$ is satisfied.

□

Basically Lemma A.10.1 states that every value v that can be constructed in s_1 can also be constructed in s_2, while preserving $ESS_{\preccurlyeq}(s_1, s_2)$ for all v. It allows to directly show the following theorem.

Lemma A.10.2 (Satisfiability (SAT) Covered). *If $s_1 \sqsubseteq s_2$ then sat $\sqsubseteq (s_1, s_2)$.*

Proof. Let c be an arbitrary linked condition of the states s_1 and s_2. According to Lemma A.10.1 if c can evaluate to b in s_1 then it can evaluate to b in s_2. So if c can evaluate to *true* in s_1 (is satisfiable in s_1), then it can also evaluate to *true* in s_2 (is satisfiable in s_2). □

Lemma A.10.3 (Path Condition Preserving). *If $s_1 \sqsubseteq s_2$ and the path conditions of both states are extended with an arbitrary linked condition c, resulting in the states $\tilde{s}_1$ and $\tilde{s}_2$, defined as*

$$\tilde{s}_1 = s_1 \text{ except } pc(\tilde{s}_1) := pc(s_1) \wedge s_1[c]$$
$$\tilde{s}_2 = s_2 \text{ except } pc(\tilde{s}_2) := pc(s_2) \wedge s_2[c]$$

then $\tilde{s}_1 \sqsubseteq \tilde{s}_2$ (the covered relation will be preserved).

Proof. This immediately follows from Lemma A.10.1, since $s_1 \sqsubseteq s_2$ by assumption. □

Lemma A.10.4 (Assignment Preserving). *If $s_1 \sqsubseteq s_2$ and the assignment $v = \mu$ is executed in both states, where μ is a linked expression, resulting in the states $\tilde{s}_1$ and $\tilde{s}_2$, defined as*

$$\tilde{s}_1 = s_1 \text{ except } vars(\tilde{s}_1) := vars(s_1)[v \leftarrow s_1[\mu]]$$
$$\tilde{s}_2 = s_2 \text{ except } vars(\tilde{s}_2) := vars(s_2)[v \leftarrow s_2[\mu]]$$

then $\tilde{s}_1 \sqsubseteq \tilde{s}_2$.

Proof. Since $s_1 \sqsubseteq s_2$ also $s_1 \sim s_2$ (by def. of $\sqsubseteq$). Thus v refers to the same variable slot in both states. So both states will modify the same variable and leave everything else unmodified. Consequently structural compatibility is preserved ($\tilde{s}_1 \sim \tilde{s}_2$), because either $s_1[\mu]$ is either completely equal to $s_2[\mu]$ or at least one of them is symbolic. According to Lemma A.10.1 $s_1 \sqsubseteq (\mu)s_2$, which then immediately implies $\tilde{s}_1 \sqsubseteq \tilde{s}_2$. □

A.10.2. Main Part

Up to this point the execution of single statements has not been considered. They have always been grouped to transitions, which are defined as a sequence of statements that are executed without interruption. However, the proof in this section requires that the single statements inside of transitions are considered. Basically the statements are handled in the same way as transitions. The execution of a statement is called a *single step*. The successor of any state s after executing a sequence of statements $l_1..l_n$ will be denoted as $\xrightarrow{l_1..l_n}$. Thus $s \xrightarrow{l_1..l_n} = \tilde{s}$ iff $s \xrightarrow{l_1..l_n} \tilde{s}$. As already defined, an error state s will be denoted as $\perp (s)$. By convention $s \xrightarrow{l} s$ iff $\perp (s)$ for all statements l. Thus an error state is never left.

Definition 38 (*Single Step Error Covered*)

A state s_1 is *single step error covered* by s_2, denoted $s_1 \sqsubseteq_\perp s_2$, iff

$$\forall l \in Statement : \left(\perp (\tilde{s}_2) \vee [\neg \perp (\tilde{s}_1) \wedge (\tilde{s}_1 \sqsubseteq \tilde{s}_2)]\right)$$

where $\tilde{s}_1 = s_1 \xrightarrow{l}$ and $\tilde{s}_2 = s_2 \xrightarrow{l}$.

Basically the above definition states that either the successor of s_2 is an error state, or the successor of s_1 is not an error state and is covered by the successor of s_2. Thus if $\perp (\tilde{s}_1)$ then $\perp (\tilde{s}_2)$ too. Single step coverage can be extended to multi step coverage. If an error state is reached from s_1 executing a path p, then there exists a prefix of p such that an error state will be reached executing that prefix from s_2. A path is a sequence of statements $l_1..l_n$. A prefix of a path is a sequence of statements $l_1..l_k$ where $k \leq n$. If $k < n$ then it is a real prefix. The length of a path p will be denoted as $|p|$.

Definition 39 (*Multi Step Error Covered*)

A state s_1 is *multi step error covered* by s_2, denoted $s_1 \sqsubseteq_\perp^* s_2$, iff

$$\forall l_1..l_n \in Statement^* : (\forall i \in \{1..n\} : \perp (\tilde{s}_2^i) \vee [\forall j \leq i : \neg \perp (\tilde{s}_1^j) \wedge (\tilde{s}_1^j \sqsubseteq \tilde{s}_2^j)])$$

where $\tilde{s}_1^i = s_1 \xrightarrow{l_1..l_i}$ and $\tilde{s}_2^i = s_2 \xrightarrow{l_1..l_i}$ for all i.

Remark. According to Definition 39 whenever the execution of a path $p = l_1..l_n$ from s_1 leads to an error state $\perp (s_1 \xrightarrow{p})$, then the execution of a prefix of p from s_2 will also lead to an error state $\perp (s_2 \xrightarrow{p})$. If $\neg \perp (s_2 \xrightarrow{q})$ where q is any real prefix of p, then $\perp (s_1 \xrightarrow{p}) \sqsubseteq_\perp (s_2 \xrightarrow{p})$ will hold. The reason that s_2 might already hit an error state when executing a real prefix of p is that s_2 can (normally) assume more values than s_1. Thus an assertion violation might occur on a real prefix of p. If all assertions are rewritten as conditional gotos, then both paths of them would be feasible in that case. Thus s_2 could always reach a corresponding error state to s_1.

Theorem A.10.5

If s_1 is covered by s_2, then s_1 is also single step covered by s_2, i.e. $s_1 \sqsubseteq s_2 \implies s_1 \sqsubseteq_\perp s_2$

Proof. The proof will consider all possible cases. Each case corresponds to a different statement. A list of valid IVL statements is available in Figure A.3. The assumption is that $s_1 \sqsubseteq s_2$ holds before the execution. After execution of a single statement t, the states $\tilde{s}_1$ and $\tilde{s}_2$ are reached. It will be shown that $\tilde{s}_1 \sqsubseteq_\perp \tilde{s}_2$ then will be valid.

The observation is, that an error state is only reached if an assertion violation is detected[7]. Thus only the *assert* statement can have influence, whether $\perp (\tilde{s}_1)$ or $\perp (\tilde{s}_2)$. So for all other statements $\neg \perp (\tilde{s}_1)$ and $\neg \perp (\tilde{2}_1)$. Thus it has to be shown for them, that they preserve the $\sqsubseteq$ relation.

Execution of any statement will modify the program counter. Either it will be incremented or in case of a goto set to the target label position. Since both states start with the same program counter, they also end up with the same program counter. So changes to the program counter (which are orthogonal to other state changes) will not invalidate the $\sqsubseteq$ relation.

Statements that do not use or modify symbolic state parts also trivially preserve the $\sqsubseteq$ relation, as they start with the same concrete state parts and apply the same operations, thus they end up again with the same concrete state parts. Among other all statements that interact with the simulation kernel belong into this category: wait_event, wait_time, notify_event, resume, suspend. All other statements will be considered in the following.

case assignment $(v = e)$

Is satisfied according to Lemma A.10.4.

case assume (assume c)

If $pc(s_1) \wedge c$ is unsatisfiable, then the execution path corresponding to s_1 terminates, that means $\tilde{s}_1$ will be a terminal state. Thus by definition of $\sqsubseteq$, it will be covered by $\tilde{s}_2$. Else $s_1' \sqsubseteq s_2'$ according to Lemma A.10.3 (extending the path condition preserves the covered relation).

case assert (assert c)

The condition $s_1[\neg c]$ can either be satisfiable $(\perp (\tilde{s}_1))$ or not $(\neg \perp (\tilde{s}_1))$.

– $\perp (\tilde{s}_1)$: In this case $\perp (\tilde{s}_2)$ too, since according to Lemma A.10.2 $sat(s_1, s_1[\neg c]) \implies sat(s_2, s_2[\neg c])$.

– $\neg \perp (\tilde{s}_1)$: In this case it is unknown whether $\perp (\tilde{s}_2)$. If $\perp (\tilde{s}_2)$, then $\tilde{s}_1 \sqsubseteq_\perp \tilde{s}_2$ (regardless whether s_1 will eventually reach an error state, an error state is reached from s_2 on a real prefix of that path). If $\neg \perp (\tilde{s}_2)$ then $\tilde{s}_1 \sqsubseteq \tilde{s}_2$ is trivially valid, since both s_1 and s_2 are not changed[8].

So in both cases $\tilde{s}_1 \sqsubseteq_\perp \tilde{s}_2$ holds.

case conditional goto (if c goto l)

According to Lemma A.10.2 $sat \sqsubseteq (s_1, s_2)$. Thus if $sat(s_1, s_1[c])$ then $sat(s_2, s_2[c])$ and if $sat(s_1, \neg s_1[c])$ then $sat(s_2, \neg s_2[c])$. So if a branch is feasible in the state s_1 then also in s_2.

If a branch is feasible in s_1 it will result in a successor state s_1', where the path condition is extended accordingly (either with c or $\neg c$). Let s_2' be the corresponding successor of

---

[7]Implicit errors, like division by zero and memory access errors can be detected by inserting assert statements at appropriate locations.

[8]Except for increasing their program counter uniformly, which does not invalidate the $\sqsubseteq$ relation.

$\langle stmt \rangle$	::=	goto id			wait $\langle expr \rangle$
		if $\langle expr \rangle$ goto id			suspend id
		$\langle expr \rangle = \langle expr \rangle$			resume id
		assume $\langle expr \rangle$			wait_time $\langle expr \rangle$
		assert $\langle expr \rangle$			wait_event $\langle expr \rangle$
		notify $\langle expr \rangle$, $\langle expr \rangle$			

Figure A.3.: Relevant statements to prove single step error coverage

s_2 then $s_1' \sqsubseteq s_2'$ due to Lemma A.10.3 (extending the path condition preserves the covered relation).

If a branch is infeasible in s_1 then the corresponding execution path will not be considered. Since it is not executed at all, it cannot reach an error state. Thus by definition it is covered by the corresponding path in s_2 regardless whether it will be continued or not.

$\square$

Remark. Every time a single statement is executed in s, the successor of s, denoted $\tilde{s}$, can be one of the following: an error state, a terminal state, one normal state or two normal states (due to symbolic branch condition).

Assuming $s_1 \sqsubseteq s_2$ and both states execute the same statement resulting in $\tilde{s}_1$ and $\tilde{s}_2$ respectively, then according to the single step error coverage Theorem A.10.5 the following properties are valid:

- If s_1 reaches an error state, then s_2 will also reach an error state.

- s_2 can reach an error state, while s_1 does not.

- If $\tilde{s}_1$ is a terminal state, then $\tilde{s}_1 \sqsubseteq \tilde{s}_2$ trivially holds, since $\tilde{s}_1$ cannot reach any error state.

- If s_1 reaches one successor state $\tilde{s}_1$, then s_2 will also reach one successor state $\tilde{s}_2$ such that $\tilde{s}_1 \sqsubseteq \tilde{s}_2$.

- If s_1 reaches two successor state $\tilde{s}_{1T}$ and $\tilde{s}_{1F}$ (due to a symbolic branch condition), then s_2 will also reach two corresponding successor states $\tilde{s}_{2T}$ and $\tilde{s}_{2F}$ such that $\tilde{s}_{1T} \sqsubseteq \tilde{s}_{2T}$ and $\tilde{s}_{1F} \sqsubseteq \tilde{s}_{2F}$.

The following theorem states, that single step error coverage can be extended to multi step error coverage.

Theorem A.10.6

> Single step coverage implies multi step coverage, i.e. $s_1 \sqsubseteq_\perp s_2 \implies s_1 \sqsubseteq_\perp^* s_2$.

Proof. Multi step error coverage can be shown by induction over the length of the path p, assuming single step error coverage. Similarly to Definition 39 the abbreviations $\tilde{s}_1^i = s_1 \xrightarrow{l_1..l_i}$ and $\tilde{s}_2^i = s_2 \xrightarrow{l_1..l_i}$ will be used for all i.

IB: $|p| = 0$

Since no statement is executed at all, $s_1 \sqsubseteq_\perp^* s_2$ holds by definition.

IS: $|p| = n + 1$ and by IH. $s_1 \sqsubseteq_\perp s_2 \implies s_1 \sqsubseteq_\perp^* s_2$ for all paths q with $|q| \leq n$.

By IH. $\forall t = l_1..l_n \in Statement^* : (\forall i \in \{1..n\} : \perp (\tilde{s_2}^i) \vee [\forall j \leq i : \neg \perp (\tilde{s_1}^j) \wedge (\tilde{s_1}^j \sqsubseteq \tilde{s_2}^j)])$, thus $\perp (\tilde{s_2}^n)$ or $(\forall j \leq n : \neg \perp (\tilde{s_1}^j) \wedge (\tilde{s_1}^j \sqsubseteq \tilde{s_2}^j))$, since these conditions are valid for all i in $\{1..n\}$. So either (1) $\perp (\tilde{s_2}^n)$ or (2) $(\forall j \leq n : \neg \perp (\tilde{s_1}^j) \wedge (\tilde{s_1}^j \sqsubseteq \tilde{s_2}^j))$. Case (2) directly implies $\neg \perp (\tilde{s_1}^n)$ and $\tilde{s_1}^n \sqsubseteq \tilde{s_2}^n$, since it must be valid for any j in $\{1..n\}$.

(1) : Since $\tilde{s_1}^n \sqsubseteq \tilde{s_2}^n$ by IH., according to Definition 38 $\tilde{s_1}^n \sqsubseteq_\perp \tilde{s_2}^n$, together they imply $s_1 \sqsubseteq_\perp^* s_2$.

(2) : $\perp (\tilde{s_2}^n)$ implies $\perp (\tilde{s_2}^{n+1})$ by definition (once an error state has been reached, it will stay an error state). Together with the IH., this implies $s_1 \sqsubseteq_\perp^* s_2$.

$\square$

Theorem A.10.7 (*Main*)

Assuming $s_1 \sqsubseteq s_2$, then for all traces w, if s_1 can reach an error state executing w then s_2 will also reach an error state executing the same trace w.

Proof. Transition and trace coverages are just special cases of multi step coverage, since every trace is a sequence of transitions, which are sequences of statements. Thus assuming $s_1 \sqsubseteq s_2$ then according to Theorem A.10.6 for all traces $w = l_1..l_n$: $\perp (s_2 \xrightarrow{w}) \vee \neg \perp (s_1 \xrightarrow{w})$, which is equivalent to $\perp (s_1 \xrightarrow{w}) \implies \perp (s_2 \xrightarrow{w})$. $\square$

The main Theorem A.10.7 directly implies the desired property $s_1 \sqsubseteq s_2 \implies s_1 \preccurlyeq s_2$.

Bibliography

[Acc12]   Accellera Systems Initiative. *SystemC*. Available at http://www.systemc.org. 2012.

[Aho+06]   Alfred V. Aho et al. *Compilers: Principles, Techniques, and Tools (2Nd Edition)*. Boston, MA, USA: Addison-Wesley Longman Publishing Co., Inc., 2006. ISBN: 0321486811.

[Amn+01]   Tobias Amnell et al. "UPPAAL - Now, Next, and Future". In: *Proceedings of the 4th Summer School on Modeling and Verification of Parallel Processes*. MOVEP '00. London, UK, UK: Springer-Verlag, 2001, pp. 99–124. ISBN: 3-540-42787-2. URL: http://dl.acm.org/citation.cfm?id=646410.692519.

[APV06]   Saswat Anand, Corina S. Păsăreanu, and Willem Visser. "Symbolic Execution with Abstract Subsumption Checking". In: *Proceedings of the 13th International Conference on Model Checking Software*. SPIN'06. Vienna, Austria: Springer-Verlag, 2006, pp. 163–181. ISBN: 3-540-33102-6, 978-3-540-33102-5. DOI: 10.1007/11691617_10. URL: http://dx.doi.org/10.1007/11691617_10.

[BH05]   Dragan Bošnački and GerardJ. Holzmann. "Improving Spin's Partial-Order Reduction for Breadth-First Search". English. In: *Model Checking Software*. Ed. by Patrice Godefroid. Vol. 3639. Lecture Notes in Computer Science. Springer Berlin Heidelberg, 2005, pp. 91–105. ISBN: 978-3-540-28195-5. DOI: 10.1007/11537328_10. URL: http://dx.doi.org/10.1007/11537328_10.

[BK10]   Nicolas Blanc and Daniel Kroening. "Race Analysis for Systemc Using Model Checking". In: *ACM Trans. Des. Autom. Electron. Syst.* 15.3 (June 2010), 21:1–21:32. ISSN: 1084-4309. DOI: 10.1145/1754405.1754406. URL: http://doi.acm.org/10.1145/1754405.1754406.

[Bla+09]   D. Black et al. *SystemC: From the Ground Up*. 2nd. Secaucus, NJ, USA: Springer-Verlag New York, Inc., 2009.

[BLL06]   Dragan Bošnački, Stefan Leue, and AlbertoLluch Lafuente. "Partial-Order Reduction for General State Exploring Algorithms". English. In: *Model Checking Software*. Ed. by Antti Valmari. Vol. 3925. Lecture Notes in Computer Science. Springer Berlin Heidelberg, 2006, pp. 271–287. ISBN: 978-3-540-33102-5. DOI: 10.1007/11691617_16. URL: http://dx.doi.org/10.1007/11691617_16.

[BMP07]   B. Bailey, G. Martin, and A. Piziali. *ESL Design and Verification: A Prescription for Electronic System Level Methodology*. Morgan Kaufmann/Elsevier, 2007.

[Bok+11]   Peter Bokor et al. "Supporting Domain-specific State Space Reductions Through Local Partial-order Reduction". In: *Proceedings of the 2011 26th IEEE/ACM International Conference on Automated Software Engineering*. ASE '11. Washington, DC, USA: IEEE Computer Society, 2011, pp. 113–122. ISBN: 978-1-4577-1638-6. DOI: 10.1109/ASE.2011.6100044. URL: http://dx.doi.org/10.1109/ASE.2011.6100044.

[CCH13]   Chun-Nan Chou, Chen-Kai Chu, and Chung-Yang (Ric) Huang. "Conquering the
            Scheduling Alternative Explosion Problem of SystemC Symbolic Simulation". In:
            Proceedings of the International Conference on Computer-Aided Design. ICCAD
            '13. San Jose, California: IEEE Press, 2013, pp. 685–690. ISBN: 978-1-4799-
            1069-4. URL: http://dl.acm.org/citation.cfm?id=2561828.2561961.

[CDE08]   Cristian Cadar, Daniel Dunbar, and Dawson Engler. "KLEE: Unassisted and Au-
            tomatic Generation of High-coverage Tests for Complex Systems Programs". In:
            *Proceedings of the 8th USENIX Conference on Operating Systems Design and
            Implementation*. OSDI'08. San Diego, California: USENIX Association, 2008,
            pp. 209–224. URL: http://dl.acm.org/citation.cfm?id=1855741.
            1855756.

[Cho+12]   Chun-Nan Chou et al. "Symbolic model checking on SystemC designs". In: *De-
            sign Automation Conference (DAC), 2012 49th ACM/EDAC/IEEE*. June 2012,
            pp. 327–333.

[Cim+10]   Alessandro Cimatti et al. "Verifying SystemC: A Software Model Checking Ap-
            proach". In: *Proceedings of the 2010 Conference on Formal Methods in Computer-
            Aided Design*. FMCAD '10. Lugano, Switzerland: FMCAD Inc, 2010, pp. 51–60.
            URL: http://dl.acm.org/citation.cfm?id=1998496.1998510.

[Cim+11]   A. Cimatti et al. "KRATOS: a software model checker for SystemC". In: *Proceed-
            ings of the 23rd international conference on Computer aided verification*. CAV'11.
            Snowbird, UT: Springer-Verlag, 2011, pp. 310–316. ISBN: 978-3-642-22109-5.
            URL: http://dl.acm.org/citation.cfm?id=2032305.2032329.

[Cla+99]   E.M. Clarke et al. "State space reduction using partial order techniques". English.
            In: *International Journal on Software Tools for Technology Transfer* 2.3 (1999),
            pp. 279–287. ISSN: 1433-2779. DOI: 10.1007/s100090050035. URL: http:
            //dx.doi.org/10.1007/s100090050035.

[CNR11]   A. Cimatti, I. Narasamdya, and M. Roveri. "Boosting lazy abstraction for systemc
            with partial order reduction". In: *Proceedings of the 17th international confer-
            ence on Tools and algorithms for the construction and analysis of systems: part
            of the joint European conferences on theory and practice of software*. TACAS'11
            / ETAPS'11. Saarbrücken, Germany: Springer-Verlag, 2011, pp. 341–356. ISBN:
            978-3-642-19834-2. URL: http://dl.acm.org/citation.cfm?id=1987389.
            1987430.

[CNR13]   Alessandro Cimatti, Iman Narasamdya, and Marco Roveri. "Software Model Check-
            ing SystemC". In: *IEEE Trans. on CAD of Integrated Circuits and Systems* 32.5
            (2013), pp. 774–787. DOI: 10.1109/TCAD.2012.2232351. URL: http://dx.
            doi.org/10.1109/TCAD.2012.2232351.

[Eck+06]   W. Ecker et al. "Specification Language for Transaction Level Assertions". In:
            *High-Level Design Validation and Test Workshop, 2006. Eleventh Annual IEEE
            International*. Nov. 2006, pp. 77–84. DOI: 10.1109/HLDVT.2006.319967.

[EEH07]   W. Ecker, V. Esen, and M. Hull. "Implementation of a Transaction Level Assertion
            Framework in SystemC". In: *Design, Automation Test in Europe Conference Exhi-
            bition, 2007. DATE '07*. Apr. 2007, pp. 1–6. DOI: 10.1109/DATE.2007.364406.

[EP10]   Sami Evangelista and Christophe Pajault. "Solving the Ignoring Problem for Par-
           tial Order Reduction". In: *Int. J. Softw. Tools Technol. Transf.* 12.2 (May 2010),
           pp. 155–170. ISSN: 1433-2779. DOI: 10.1007/s10009-010-0137-y. URL:
           http://dx.doi.org/10.1007/s10009-010-0137-y.

[FG05]   Cormac Flanagan and Patrice Godefroid. "Dynamic Partial-order Reduction for
           Model Checking Software". In: *Proceedings of the 32Nd ACM SIGPLAN-SIGACT
           Symposium on Principles of Programming Languages*. POPL '05. Long Beach,
           California, USA: ACM, 2005, pp. 110–121. ISBN: 1-58113-830-X. DOI: 10.
           1145/1040305.1040315. URL: http://doi.acm.org/10.1145/1040305.
           1040315.

[Fos09]   Harry Foster. "Applied Assertion-Based Verification: An Industry Perspective".
           In: *Found. Trends Electron. Des. Autom.* 3.1 (Jan. 2009), pp. 1–95. ISSN: 1551-
           3939. DOI: 10.1561/1000000013. URL: http://dx.doi.org/10.1561/
           1000000013.

[FP09]   L. Ferro and L. Pierre. "ISIS: Runtime verification of TLM platforms". In: *Speci-
           fication Design Languages, 2009. FDL 2009. Forum on.* Sept. 2009, pp. 1–6.

[GD10]   D. Große and R. Drechsler. *Quality-Driven SystemC Design*. Springer, 2010.

[GLD10]   D. Große, H. Le, and R. Drechsler. "Proving transaction and system-level proper-
           ties of untimed SystemC TLM designs". In: *MEMOCODE*. 2010, pp. 113–122.

[God91]   Patrice Godefroid. "Using Partial Orders to Improve Automatic Verification Meth-
           ods". In: *Proceedings of the 2Nd International Workshop on Computer Aided Ver-
           ification*. CAV '90. London, UK, UK: Springer-Verlag, 1991, pp. 176–185. ISBN:
           3-540-54477-1. URL: http://dl.acm.org/citation.cfm?id=647759.
           735044.

[God96]   P. Godefroid. *Partial-Order Methods for the Verification of Concurrent Systems:
           An Approach to the State-Explosion Problem*. Ed. by J. van Leeuwen, J. Hart-
           manis, and G. Goos. Secaucus, NJ, USA: Springer-Verlag New York, Inc., 1996.
           ISBN: 3540607617.

[GP93]   Patrice Godefroid and Didier Pirottin. "Refining Dependencies Improves Partial-
           Order Verification Methods (Extended Abstract)". In: *Proceedings of the 5th In-
           ternational Conference on Computer Aided Verification*. CAV '93. London, UK,
           UK: Springer-Verlag, 1993, pp. 438–449. ISBN: 3-540-56922-7. URL: http://
           dl.acm.org/citation.cfm?id=647762.735513.

[Gro02]   T. Grotker. *System Design with SystemC*. Norwell, MA, USA: Kluwer Academic
           Publishers, 2002. ISBN: 1402070721.

[GW93]   Patrice Godefroid and Pierre Wolper. "Using Partial Orders for the Efficient Verifi-
           cation of Deadlock Freedom and Safety Properties". In: *Form. Methods Syst. Des.*
           2.2 (Apr. 1993), pp. 149–164. ISSN: 0925-9856. DOI: 10.1007/BF01383879.
           URL: http://dx.doi.org/10.1007/BF01383879.

[Hae+11]   F. Haedicke et al. "metaSMT: Focus on Your Application not on Solver Integra-
           tion". In: *DIFTS*. 2011, pp. 22–29.

[Hel+06]   C. Helmstetter et al. "Automatic Generation of Schedulings for Improving the
           Test Coverage of Systems-on-a-Chip". In: *Proceedings of the Formal Methods in
           Computer Aided Design*. FMCAD '06. Washington, DC, USA: IEEE Computer
           Society, 2006, pp. 171–178. ISBN: 0-7695-2707-8. DOI: 10.1109/FMCAD.2006.
           10. URL: http://dx.doi.org/10.1109/FMCAD.2006.10.

[HFG08]   Paula Herber, Joachim Fellmuth, and Sabine Glesner. "Model Checking SystemC
           Designs Using Timed Automata". In: *Proceedings of the 6th IEEE/ACM/IFIP In-
           ternational Conference on Hardware/Software Codesign and System Synthesis*.
           CODES+ISSS '08. Atlanta, GA, USA: ACM, 2008, pp. 131–136. ISBN: 978-1-
           60558-470-6. DOI: 10.1145/1450135.1450166. URL: http://doi.acm.org/
           10.1145/1450135.1450166.

[HGP92]   Gerard J. Holzmann, Patrice Godefroid, and Didier Pirottin. "Coverage Preserv-
           ing Reduction Strategies for Reachability Analysis". In: *Proceedings of the IFIP
           TC6/WG6.1 Twelfth International Symposium on Protocol Specification, Testing
           and Verification XII*. Amsterdam, The Netherlands: North-Holland Publishing Co.,
           1992, pp. 349–363. ISBN: 0-444-89874-3. URL: http://dl.acm.org/citation.
           cfm?id=645835.670549.

[HMM09]   C. Helmstetter, F. Maraninchi, and L. Maillet-Contoz. "Full simulation coverage
           for SystemC transaction-level models of systems-on-a-chip". English. In: *Formal
           Methods in System Design* 35.2 (2009), pp. 152–189. ISSN: 0925-9856. DOI: 10.
           1007/s10703-009-0075-z. URL: http://dx.doi.org/10.1007/s10703-
           009-0075-z.

[IEE05]   IEEE. *IEEE Standard for Property Specification Language (PSL)*. IEEE Std. 1850-
           2005. 2005.

[IEE11]   IEEE. *IEEE Standard SystemC Language Reference Manual*. IEEE Std. 1666.
           2011.

[Ios02]   Radu Iosif. "Symmetry Reduction Criteria for Software Model Checking". In:
           *Proceedings of the 9th International SPIN Workshop on Model Checking of Soft-
           ware*. London, UK, UK: Springer-Verlag, 2002, pp. 22–41. ISBN: 3-540-43477-1.
           URL: http://dl.acm.org/citation.cfm?id=645881.672231.

[KEP06]   Daniel Karlsson, Petru Eles, and Zebo Peng. "Formal Verification of Systemc De-
           signs Using a Petri-net Based Representation". In: *Proceedings of the Conference
           on Design, Automation and Test in Europe: Proceedings*. DATE '06. Munich, Ger-
           many: European Design and Automation Association, 2006, pp. 1228–1233. ISBN:
           3-9810801-0-6. URL: http://dl.acm.org/citation.cfm?id=1131481.
           1131824.

[KGG08]   Sudipta Kundu, Malay Ganai, and Rajesh Gupta. "Partial Order Reduction for
           Scalable Testing of systemC TLM Designs". In: *Proceedings of the 45th An-
           nual Design Automation Conference*. DAC '08. Anaheim, California: ACM, 2008,
           pp. 936–941. ISBN: 978-1-60558-115-6. DOI: 10.1145/1391469.1391706. URL:
           http://doi.acm.org/10.1145/1391469.1391706.

[Kin76]   James C. King. "Symbolic Execution and Program Testing". In: *Commun. ACM*
           19.7 (July 1976), pp. 385–394. ISSN: 0001-0782. DOI: 10.1145/360248.360252.
           URL: http://doi.acm.org/10.1145/360248.360252.

[KS05]   D. Kroening and N. Sharygina. "Formal Verification of SystemC by Automatic Hardware/Software Partitioning". In: *Proceedings of the 2Nd ACM/IEEE International Conference on Formal Methods and Models for Co-Design*. MEMOCODE '05. Washington, DC, USA: IEEE Computer Society, 2005, pp. 101–110. ISBN: 0-7803-9227-2. DOI: 10.1109/MEMCOD.2005.1487900. URL: http://dx.doi. org/10.1109/MEMCOD.2005.1487900.

[KWG09]   Vineet Kahlon, Chao Wang, and Aarti Gupta. "Monotonic Partial Order Reduction: An Optimal Symbolic Partial Order Reduction Technique". In: *Proceedings of the 21st International Conference on Computer Aided Verification*. CAV '09. Grenoble, France: Springer-Verlag, 2009, pp. 398–413. ISBN: 978-3-642-02657-7. DOI: 10.1007/978-3-642-02658-4_31. URL: http://dx.doi.org/10. 1007/978-3-642-02658-4_31.

[Le+13]   Hoang M. Le et al. "Verifying SystemC Using an Intermediate Verification Language and Symbolic Simulation". In: *Proceedings of the 50th Annual Design Automation Conference*. DAC '13. Austin, Texas: ACM, 2013, 116:1–116:6. ISBN: 978-1-4503-2071-9. DOI: 10.1145/2463209.2488877. URL: http://doi.acm. org/10.1145/2463209.2488877.

[Maz87]   A Mazurkiewicz. "Trace Theory". In: *Advances in Petri Nets 1986, Part II on Petri Nets: Applications and Relationships to Other Models of Concurrency*. Bad Honnef: Springer-Verlag New York, Inc., 1987, pp. 279–324. ISBN: 0-387-17906-2. URL: http://dl.acm.org/citation.cfm?id=25542.25553.

[MMM05]   Matthieu Moy, Florence Maraninchi, and Laurent Maillet-Contoz. "LuSsy: An open tool for the analysis of systems-on-a-chip at the transaction level". English. In: *Design Automation for Embedded Systems* 10.2-3 (2005), pp. 73–104. ISSN: 0929-5585. DOI: 10.1007/s10617-006-9044-6. URL: http://dx.doi.org/ 10.1007/s10617-006-9044-6.

[Pel93]   Doron Peled. "All from one, one for all: on model checking using representatives". English. In: *Computer Aided Verification*. Ed. by Costas Courcoubetis. Vol. 697. Lecture Notes in Computer Science. Springer Berlin Heidelberg, 1993, pp. 409–423. ISBN: 978-3-540-56922-0. DOI: 10.1007/3-540-56922-7_34. URL: http://dx.doi.org/10.1007/3-540-56922-7_34.

[PF08]   L. Pierre and L. Ferro. "A Tractable and Fast Method for Monitoring SystemC TLM Specifications". In: *Computers, IEEE Transactions on* 57.10 (Oct. 2008), pp. 1346–1356. ISSN: 0018-9340. DOI: 10.1109/TC.2008.74.

[Sim+13]   Jiri Simsa et al. "Scalable Dynamic Partial Order Reduction". In: *In Proceedings of the 3rd International Conference on Runtime Verifications, RV'12*. 2013.

[Tab+08]   Deian Tabakov et al. "A Temporal Language for SystemC". In: *Proceedings of the 2008 International Conference on Formal Methods in Computer-Aided Design*. FMCAD '08. Portland, Oregon: IEEE Press, 2008, 22:1–22:9. ISBN: 978-1-4244-2735-2. URL: http://dl.acm.org/citation.cfm?id=1517424.1517446.

[TC11]   Linda Torczon and Keith Cooper. *Engineering A Compiler*. 2nd. San Francisco, CA, USA: Morgan Kaufmann Publishers Inc., 2011. ISBN: 012088478X.

[Tra+07]   Claus Traulsen et al. "A systemC/TLM Semantics in PROMELA and Its Possi-
            ble Applications". In: *Proceedings of the 14th International SPIN Conference on
            Model Checking Software*. Berlin, Germany: Springer-Verlag, 2007, pp. 204–222.
            ISBN: 978-3-540-73369-0. URL: http://dl.acm.org/citation.cfm?id=
            1770532.1770552.

[TV10]   D. Tabakov and M.Y. Vardi. "Monitoring temporal SystemC properties". In: *For-
            mal Methods and Models for Codesign (MEMOCODE), 2010 8th IEEE/ACM In-
            ternational Conference on*. July 2010, pp. 123–132. DOI: 10.1109/MEMCOD.
            2010.5558640.

[Val89]   Antti Valmari. "Stubborn Sets for Reduced State Space Generation". In: *Proceed-
            ings of the Tenth International Conference on Application and Theory of Petri
            Nets*. 1989, pp. 1–22.

[Val91]   Antti Valmari. "A Stubborn Attack On State Explosion". In: *Proceedings of the
            2Nd International Workshop on Computer Aided Verification*. CAV '90. London,
            UK, UK: Springer-Verlag, 1991, pp. 156–165. ISBN: 3-540-54477-1. URL: http:
            //dl.acm.org/citation.cfm?id=647759.735025.

[Val98]   Antti Valmari. "The state explosion problem". English. In: *Lectures on Petri Nets
            I: Basic Models*. Ed. by Wolfgang Reisig and Grzegorz Rozenberg. Vol. 1491.
            Lecture Notes in Computer Science. Springer Berlin Heidelberg, 1998, pp. 429–
            528. ISBN: 978-3-540-65306-6. DOI: 10.1007/3-540-65306-6_21. URL:
            http://dx.doi.org/10.1007/3-540-65306-6_21.

[Var07]   Moshe Y. Vardi. "Formal Techniques for SystemC Verification". In: *Proceedings
            of the 44th Annual Design Automation Conference*. DAC '07. San Diego, Cal-
            ifornia: ACM, 2007, pp. 188–192. ISBN: 978-1-59593-627-1. DOI: 10.1145/
            1278480.1278527. URL: http://doi.acm.org/10.1145/1278480.1278527.

[VV98]   Kimmo Varpaaniemi and Kimmo Varpaaniemi. *On the Stubborn Set Method in
            Reduced State Space Generation*. Tech. rep. 1998.

[Yan+07]   Yu Yang et al. "Distributed Dynamic Partial Order Reduction based Verification of
            Threaded Software". In: *In Workshop on Model Checking Software (SPIN 2007*.
            2007.

[Yan+08]   Yu Yang et al. "Efficient Stateful Dynamic Partial Order Reduction". In: *Proceed-
            ings of the 15th International Workshop on Model Checking Software*. SPIN '08.
            Los Angeles, CA, USA: Springer-Verlag, 2008, pp. 288–305. ISBN: 978-3-540-
            85113-4. DOI: 10.1007/978-3-540-85114-1_20. URL: http://dx.doi.org/
            10.1007/978-3-540-85114-1_20.

[YWY06]   Xiaodong Yi, Ji Wang, and Xuejun Yang. "Stateful Dynamic Partial-Order Re-
            duction". English. In: *Formal Methods and Software Engineering*. Ed. by Zhim-
            ing Liu and Jifeng He. Vol. 4260. Lecture Notes in Computer Science. Springer
            Berlin Heidelberg, 2006, pp. 149–167. ISBN: 978-3-540-47460-9. DOI: 10.1007/
            11901433_9. URL: http://dx.doi.org/10.1007/11901433_9.

Printed in the United States
By Bookmasters